M&E TECBOOKS

The primary need of students taking Technician ation Council (TEC) courses is for books which ect the new methodology and syllabus requirements. In presenting M & E TECBOOKS, we believe that this careful collaboration between subject editors and authors has resulted in the very best, "tailor-made" texts which could be devised for the student – and which are likely to be of guidance to the lecturer also. The aim has been to provide books based on TEC's own objectives, concisely and authoritatively presented, priced as closely as possible to the student budget.

D0815646

General Editor

Dr Edwin Kerr
Chief Officer, Council for National Academic Awards

Subject Editors

D. Anderson
*Head of Department of General
Building, Leeds College of Building*

Dr F. Goodall
*Head of Department of Engineering,
Salford College of Technology*

L. G. North
*Head of Department of Biological Sciences,
North East Surrey College of Technology*

C. J. Thompson
*Head of Department of Science,
Matthew Boulton Technical College*

. Ashen
*Head of Department of Hotel and Catering Science,
Plymout   lege of Further Education*

G. A. Woolvet
*He   School of Mechanical,
Aeronau   and Production Engineering,
ston Polytechnic*

The M&E TECBOOK Series

Structural Detailing Level II

BRIAN CURRIE
C Eng MIStructE

*Senior Lecturer in Reinforced Concrete Design and Detailing,
Ulster Polytechnic*

ROBERT A SHARPE
C Eng MIStructE MIPHE

*Senior Lecturer in Structural Steelwork Design and Detailing,
Ulster Polytechnic*

Macdonald and Evans

Macdonald & Evans Ltd
Estover, Plymouth PL6 7PZ

First published 1982

© Macdonald & Evans Ltd 1982

ISBN: 0 7121 1985 X

Filmset in Monophoto Times by
Northumberland Press Ltd, Gateshead
Printed in Great Britain by
Richard Clay (The Chaucer Press) Ltd,
Bungay, Suffolk

Preface

The authors have felt for some time the need for a book which combines both the detailing of reinforced concrete and structural steelwork. This information has been found in various scattered sources, and a lot has been handed down from preceding generations in the structural drawing office. When TEC produced standard units in structural detailing it was thought imperative that a book designed specifically to meet this need should be written.

The purpose of this book is to give the student a text closely related to the requirements of the TEC syllabus for Structural Detailing II. At the close of each chapter, self-assessment questions are included to enable the student to test his grasp of the contents of that particular chapter.

Where deemed necessary the content of the text has been enlarged to satisfy the needs of a wider readership, and the authors would be disappointed if this book did not prove valuable to practising engineers as well as TEC students and undergraduates.

The authors between them, have over thirty years' experience in the field of design and detailing of structures, and are in no doubt that this experience has been of major value in the production of this volume.

While a passing need is met by this book, the authors have a deep conviction that the student should be aware of something of much more abiding importance. This is aptly summarised in the words of G. E. Troy:

"What will it profit when life here is o'er,
Though great worldly wisdom I gain,
If, seeking knowledge, I utterly fail
The wisdom of God to obtain"

(I Corinthians i, 17–25)

September 1982   BC
                                                 RAS

Acknowledgments

The authors wish to thank all those who have rendered valuable assistance during the preparation of this book. It would be impossible to mention each individual by name but special reference should be made to the following.

Professor R. B. Schofield, Director of Studies of the School of Civil Engineering, Ulster Polytechnic, who has given every encouragement to this venture. Mr D. J. O'Connor for his helpful and constructive criticism.

To many societies, institutions and bodies who gave permission to reproduce material. Extracts have been taken from:

Standard Method of Detailing Reinforced Concrete published for The Concrete Society and the Institution of Structural Engineers by the Cement and Concrete Association; *Standard Reinforced Concrete Details* published for the Concrete Society by the Cement and Concrete Association; CP110 *The Structural Use of Concrete*, BS4466 *Bending Dimensions and Scheduling of Bars for the Reinforcement of Concrete*, BS449 *The Use of Structural Steel in Building*, and BS499: Part 2: 1980 *Welding Symbols*, all published by the British Standards Institution, from whom copies of the complete standards may be purchased; *Structural Steelwork Handbook* published by The British Constructional Steelwork Association Limited, and the Constructional Steel Research and Development Organisation; *Metric Practice for Structural Steelwork* and *Structural Fasteners and their Application* both published by the British Constructional Steelwork Association Limited.

We must convey our thanks to Mrs Rea, Mrs McCauley, Mrs Clarke and Mrs Convery for typing the manuscript so efficiently, and finally to our respective families for their patience and encouragement during the time of script preparation.

Contents

The Reason Why

CHAPTER OBJECTIVES

After studying this chapter you should be able to:
* understand the reasons for producing structural drawings, together with the other necessary information;
* understand how these documents assist the related professions.

WHY HAVE DRAWINGS?

The number of professions, trades, and hence people involved, from the client stage of a contract to its completion is legion. Directly involved are:

(*a*) the client;
(*b*) architect;
(*c*) structural engineer;
(*d*) quantity surveyor;
(*e*) mechanical engineer;
(*f*) electrical engineer;
(*g*) main contractor;
(*h*) and a host of sub-contractors.

Indirectly involved are the planning and building control authorities. The proposals and requirements can be best communicated to these many and varied people by drawings.

It is helpful for the student to understand the procedure usually adopted during the course of unravelling this complex problem. This can best be outlined by considering the role played by those involved in the process and the various duties that each must perform.

The client and the architect
First we have the client, who is the instigator or source of a contract. He provides the original brief and the finance for the work. The first step he takes is usually to approach an architect, and after some discussion this leads to a firm appoint-ment. The architect becomes responsible for the production of a building which satisfies the client both in terms of serviceability and finance whilst being aesthetically acceptable to both the client and the public. This latter aspect forms an important part of the responsibilities of the planning authority.

Conception
The appearance of the building is unknown at this early stage of discussion between client and architect. There are, however, various parameters such as the height and architecture of adjacent buildings, the floor area required by the client, and, not least, finance, which guide the thinking of the architect. Applying experience, flair and good judgment to these parameters the appearance of the building is conceived in the mind of the architect. This is sketched by the architect in order to convey his thoughts to the client, who may or may not like the suggestion; however, after some discussion and perhaps rethinking, a scheme is eventually agreed between them. Once this has been done a start can be made on the detailed design, costing, etc., for the project.

Calculation
At some stage between his appointment and the client approving a particular scheme the architect will appoint the other members of the professional team—the structural engineer, quantity surveyor and services (mechanical, heating and ventilating, and electrical) engineer. It is essential for these people to be introduced early in this conceptual stage of the design in order to assess the feasibility of the architectural proposals.

The structural engineer and the structural detailer
As far as the structural engineer is concerned he

will use his experience and ingenuity rather than precise calculation to guide the architect in the very early days. However, once the architect and client have agreed on the scheme, the process of producing calculations and drawings for the structure of the building will begin. The position of the supporting columns and walls, and hence the beams, will have been decided in consultation with the other members of the design team and these, in the first instance, are indicated on the architectural layout drawings. The structural engineer must often calculate the necessary sizes and characteristics of these structural members. Once this has been done it is the responsibility of the structural detailer to produce drawings showing clearly these sizes and characteristics. He will use the calculations and the architectural drawings to produce general arrangement drawings showing the layout, typical floor and roof plans and sections through the building, all setting out clearly and unambiguously the proposed structural scheme. From these the structure is broken down into its various elements which are isolated, drawn and detailed.

The structural engineer will usually calculate the loads and thus size the members, beginning at roof level and working down to the foundations. When, however, the detail design and drawings are being produced the reverse order is adopted, working from foundation to roof level, i.e. in the sequence of construction.

Experience has shown the value of covering certain essential factors in a thorough and orderly manner and of incorporating these approaches in codes of practice. A number of these exist and a few are listed below. They are explained later in the text.

CP3, Chapter V, *Dead, Imposed and Wind Loads*.

CP110, *The Structural Use of Concrete*.

BS449, *The Use of Structural Steel in Buildings*.

The quantity surveyor

It is essential that careful financial control is kept at all stages of the contract. This is the job of the quantity surveyor (QS). He is involved at an early stage in the contract and can comment, from a financial point of view, on the merits of different building materials. He also puts restraints on various members of the team to ensure that the finished contract price is within the original budget of the client.

When the drawings of both the architect and the structural engineer are sufficiently advanced the quantity surveyor extracts from these the information required to produce a document called a bill of quantities. This lists, in a standard format, each part of the work to be done and the materials required, leaving blank the cost of labour and materials. Upon completion of the bill various contractors price the job on these prepared sheets, yielding an overall price for the job. This is known as tendering for a contract and the job is said to be "out to tender".

When these tenders are complete and returned a main contractor is appointed. The quantity surveyor will measure the work as it is done on site, compare with the relevant drawings, noting all revisions, and pay the contractor accordingly while keeping a watchful eye on the overall financial position.

COMMUNICATION

All that has been written thus far describing the various stages to be gone through to produce a successful job could be summarised in one word —communication. We all spend our lives communicating, sometimes orally, sometimes in writing, and occasionally by signs. It has been a feature of man from his earliest times to communicate in a pictorial fashion. We should not be surprised, therefore, to find that the art of drawing has been developed to such a degree that the whole of the civil engineering and building industry depends to a large extent on successful communication by drawing.

The line of communication has been traced from the client to the architect, structural engineer and quantity surveyor, but there are other organisations and people involved in the various stages leading up to the successful completion of a contract. They are many and varied and include consultants, government departments, contractors, specialist firms, tradesmen on site, and so on. Their relationships to and involvement in the contract are described below.

Statutory bodies

Approval for all work must be obtained from the planning and building control authorities prior to any work commencing. It is usual to initiate discussions with these bodies at an early stage in the proceedings. This is normally done at the design stage of the scheme, to minimise any abortive work. Drawings and calculations are submitted for approval showing the structural stability of the work at all stages of construction and highlighting any unusual features.

Services engineer

Each building must have heat, light, ventilation, water, drainage and sewerage services. With the exception of internal storm-water drainage, usually dealt with by the architect, these other matters are in the sphere of the services engineer. Again, the earlier he is involved the better. Obviously the structural engineer needs to know the location of plant rooms, vertical service shafts and other holes through the structure to cater for the distribution of the services. This information is obtained from discussions between the architect, the structural engineer and the services engineer. As the detail drawings for the floors proceed from the structural engineer copies are sent to the services engineer in order that the size and position of any necessary holes can be indicated. This is done prior to the detailing of the reinforcement. However, due to the fact that it is the services contractor and not the consultant who produces the working drawings for the services, this is an area where a serious breakdown in communications can occur.

Sub-contractors

The structural engineer has to provide information not only for the main contractor but also for a number of sub-contractors, the two principal ones being the piling and the structural steelwork sub-contractors.

The piling sub-contractor

He will receive from the structural engineer a copy of the proposed foundation layout, together with the design working loads. Using this information and the site investigation report, the piling sub-contractor produces his design and after further discussion a layout is approved. This process would normally be instigated by putting the work out to tender, but sometimes it may be "negotiated" if, for example, there are few specialist firms available to carry out this work.

The steelwork sub-contractor

The role of the steelwork sub-contractor demands more consideration. In a simple job the architect may approach a steel firm directly without a consultant engineer, but the more usual procedure adopted involves the structural consultant designing the structural steelwork and producing steel layout drawings showing the sizes of the various members.

If the consultant is not designing the connections he must indicate the reactions and, if appropriate, the shearing forces and bending moments arising at each connection. Again this can be the basis for the appointment of the steelwork sub-contractor either by negotiation or tender.

Work in the drawing office. After appointment the steelwork sub-contractor will prepare his own drawings and schedules mainly for his internal use. In the drawing office the detailer produces general arrangement drawings showing the steel layout, including erection marks and end reactions, direction of span of floor slabs and leading dimensions. One of the most important tasks for the detailer is the preparation of drawings for the fabrication shop. These detail fully the size of section required, number of elements, exact cutting length, plate and angle cleat connections, drilling of holes, and the identification or erection mark to be painted on each member corresponding to that shown on the general arrangement drawings.

Work in the fabricating shop. In addition to those tradesmen engaged in cutting, drilling and welding, an important person in the fabricating shop is the template maker or loftsman. He uses timber laths or battens and special paper to make full-size templates or patterns for all members and connections in a lattice framework. However for simple beam/stanchion junctions, templates are produced for connections only. On each template the diameter of holes and an identification mark are indicated for cross-referencing with detail drawings and for fabrication purposes. When all is seen to fit in an acceptable fashion the steelwork is fabricated, identification marks are painted on after any other specialised painting has been done, and all necessary bolts for site connections prepared, all ready for delivery to site.

Other sub-contractors associated with the steelwork

These steelwork drawings are also used by the sheeting sub-contractors to determine the lengths and quantities of roof, side sheets and fixing bolts to match the spacing of the roof purlins and the sheeting rails. Quantities of sheets, ridge capping, eaves fillers, gutters and downpipes are also extracted from these drawings. Patent glazing sub-contractors use the drawings to extract sizes and quantities of glazing bars, glass and lead flashings.

Again, the floor suppliers will find the steel drawings of great help. Whether of precast concrete or timber, the length of members, type and position of connection can be determined to suit the structural frame. In the case of an in situ concrete floor the reinforcement must be detailed to suit the relative position and levels of the steel floor beams and the concrete floor.

Where fire protection to the steel frame is provided by sheeting, e.g. asbestos, the supplier uses these drawings to determine the quantity required.

After all has been co-ordinated by the architect the various parts of the scheme should fit together like a large jigsaw puzzle. This is done on the site and the person responsible for doing it is the main contractor. Thus the line of communication for the construction process ends with the contractor.

If you wish to pursue this topic of communication further it is recommended that you read a report by the Institution of Structural Engineers entitled *Communication of Structural Design*, published in April 1975, and an ensuing discussion in the March 1976 edition of *The Structural Engineer*.

The main contractor

After the tendering stage outlined above, the main contractor for the project is appointed. When he arrives on site the contractor will be armed with sufficient drawings to enable him to grasp the shape, size and appearance of the building, how to set it out accurately and, as far as the structure is concerned, the overall structural scheme in the form of general arrangement drawings. For a reinforced concrete job he would expect to have the structural details of the foundations, the ground floor and perhaps the columns and walls to first floor level, whereas for a structural steelwork job he will have complete details from the steelwork sub-contractor. This implies that where the structure is of concrete reinforced in situ, there is a time during the contract when the structural engineering and detailing staff and the contracting staff are working in parallel, the office staff being sufficiently far in advance of the contractor to enable him to order and take delivery of the necessary materials without causing any delay to the contract. The co-ordination of this work must be carried out carefully by the structural engineer and the contractor if delays in the contract are to be avoided.

In the case of a structural steelwork contract the construction may be undertaken by the main contractor, but it is more usual for the steelwork sub-contractor to undertake his own erection. The main contractor is usually supplied with a foundation plan for the structural steel sub-contract showing the position and details of the holding-down bolts and washer plates to be incorporated in the concrete foundation. These provide the necessary fixings for the steel stanchions.

The steel erector then constructs the steel frame in accordance with the drawings, although if protection to the frame is of in situ concrete this is very often undertaken by the main contractor upon completion of the steelwork erection.

Just as the architect is responsible for planning and co-ordinating the office side of the work, so the main contractor has similar responsibilities on site. Thus he must be aware of any sub-contracting work which is on the critical path of the contract.

Conclusion

It is generally agreed that information is most quickly absorbed when it enters the mind by way of the "eye-gate". Thus drawings utilise a language which is capable of overcoming barriers of educational background, culture, age and even nationality, enabling a building envisaged in the mind of one person to become a working reality. This can only be achieved by careful, clear and unambiguous drawings and this is the realm and responsibility of the structural detailer. His work becomes the real line of communication between the structural engineer and the contractor and no amount of discussion on site is a substitute for good drawings. How these are produced is the subject of Chapter Two.

WHY HAVE SCHEDULES?

A schedule is essentially a list of materials which enables ordering to take place. Schedules generally supplement drawings and can refer to a variety of components. Examples are door and window schedules, reinforcing bar-bending schedules, and so on.

We shall now consider the usefulness of schedules in more detail. You will readily appreciate the need for a structure of some description to withstand the internal and external loads imposed upon a building. This structure can be constructed from a single material or more commonly from a combination of many different materials. A simple example is that of a house where the foundations are of concrete, the walls are of brickwork or blockwork or both, the lintels are of steel and/or concrete and the floor and roof made of timber. In such a situation the main fabric can be measured and ordered directly from a drawing, e.g. the number of bricks required for a wall can be calculated when the elevational area to be covered is known in square metres (approximately 60 bricks per square metre per 100 mm skin). However for the more complex structures the drawings need to be supplemented. The architect, for example, will supply door and window schedules which are needed by a contractor to ensure the correct ordering of materials. The structural engineer is in a similar situation since, in the case of reinforced concrete, the drawings require supplementation in the form of reinforcement schedules; if structural steelwork is the medium used

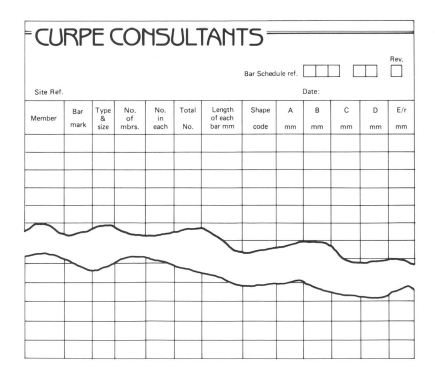

Fig. 1. *A bar bending schedule sheet.* For the layout dimensions for this sheet and also the schedule sheet for fabric see Figs. 2 and 3 in BS4466 : 1981.

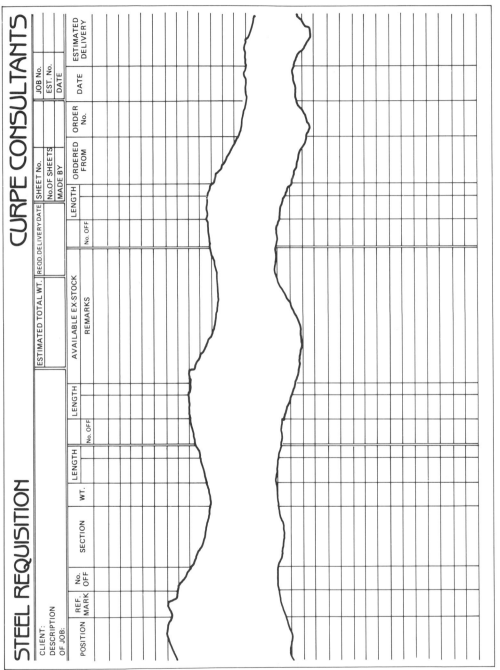

Fig. 2. *A steel requisition list.*

the related information is in the form of materials and bolt lists.

Having shown that there is a need for schedules, we now describe the relevant types and indicate their uses.

Bar-bending schedules

When a reinforced concrete structure is being drawn, each element of the structure is isolated and detailed; thus there are separate drawings for beams, columns, walls, slabs and foundations. Enough information must be given on these drawings to allow the profile to be clearly grasped and constructed. It is also necessary to show exactly what reinforcement is required and where this should be fixed. Each reinforcing bar must be cut and bent so that it fits exactly into place when delivered to the site. There must then be a means of listing each bar, showing the type of reinforcement, i.e. whether high-yield steel or mild steel, the number of bars required, their length, the shape required and the bending dimensions. A typical bar-bending schedule sheet is shown in Fig. 1. This sheet is completed by the reinforced concrete detailer in the struc-

tural consultant's office, with the result that every reinforcing bar on the job has a unique specification; this should rule out the possibility of ambiguity.

Material lists

The structural steelwork sub-contractor uses the steel layout drawings to prepare the first of two steel requisition lists; this shows the material required for main structural members and is completed prior to detailing. When the detail drawings are complete the second list is produced, indicating the material necessary for the connections. Each of these lists shows the section size required, number off, and approximate lengths. A typical sheet is shown in Fig. 2.

When the detail drawings are completed and checked, material cutting lists are produced in the drawing office, giving detailed information for members and connections, e.g. identification mark, section size, number off, exact cutting length. A typical sheet is shown in Fig. 3.

Bolt lists

The elements of a steel frame are fabricated in

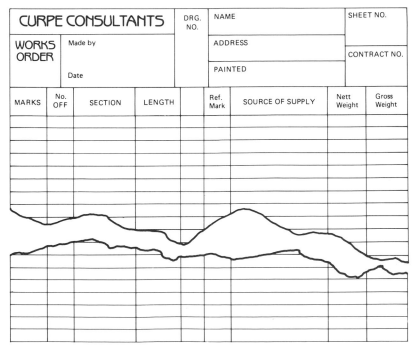

Fig. 3. *A material cutting list.*

shops, where a high degree of control is possible, painted as required and assembled for transporting to the site. The detailer will have broken down the frame into sizes suitable for the proposed methods of transport and erection. It is apparent that some bolts will be used for shop assembly (shop bolts) and others will be required for connections on the site (site bolts).

Two types of bolts are most commonly used, viz. black bolts and high strength friction grip bolts. Black bolts are normally used but high strength friction grip (HSFG) bolts are ideal for rigid joints such as the knee and apex joints in portal frame construction.

The length of bolts can be determined from the thickness of the elements to be connected, plus an allowance for the washer, nut and projection.

Bolt lists are made out for shop bolts and site bolts, stating the number required, type, diameter, length and location. Holding down (HD) bolts and plate washers used for fixing steel stanchions to concrete foundations have a separate list as they are required by the main contractor on site before the foundations are poured. A typical sheet is shown in Fig. 4.

Who uses the bar bending schedules?
The importance of the bar-bending schedule as supplementary information may be gauged from a consideration of the varied and numerous people who use them. The report of a joint committee from the Concrete Society and the Institution of Structural Engineers gives the following list:

(*a*) the detailer;
(*b*) the person checking the drawing;
(*c*) the contractor who orders the reinforcement;
(*d*) the organisation responsible for fabricating the reinforcement;

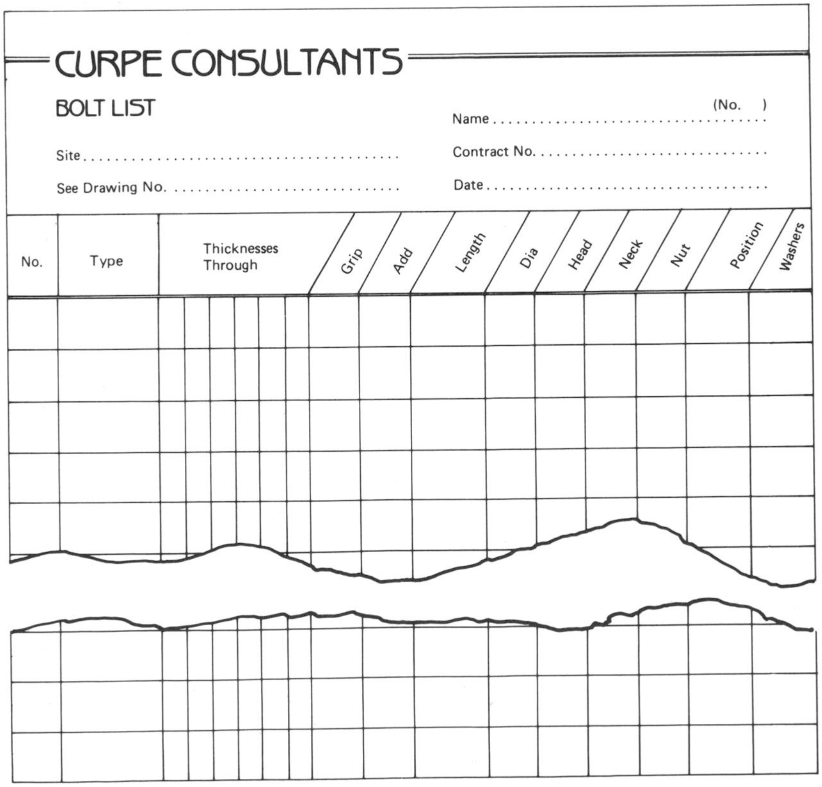
CURPE CONSULTANTS
BOLT LIST
(No.)
Name ....
Site ....
Contract No. ....
See Drawing No. ....
Date ....

No.	Type	Thicknesses Through	Grip	Add	Length	Dia	Head	Neck	Nut	Position	Washers

Fig. 4. *A bolt list.*

(*e*) the steel fixer;
(*f*) the clerk of works or other inspector;
(*g*) the quantity surveyor.

Who uses the materials list and the bolt list?
Again their importance is realised by considering the number of people who use them:

(*a*) the detailer;
(*b*) the checker;
(*c*) the storeman;
(*d*) the steel stockholder;
(*e*) the rolling mill;
(*f*) the fabricator;
(*g*) the erector;
(*h*) the quantity surveyor or estimator.

What is a bar-bending schedule used for?
When a bar-bending schedule is completed copies are sent to various people who use them for different purposes.

The quantity surveyor will receive a copy to enable him to keep his costs up to date. It also allows him to keep a check on the original reinforcement estimate, prepared by the structural engineer, which formed the basis of the contract price. This comparison provides very useful feedback for the structural engineer.

The contractor will get a number of copies, one of which he will forward to the reinforcement supplier who will cut and bend the reinforcement to the specified dimensions. Another he will give to his own quantity surveyor who also keeps a check on the financial state of the contract. When the time comes for a particular element of structure to be produced the steel fixer will receive a copy, along with the relevant drawing, to ensure that the correct reinforcement is placed in the proper position.

Finally the clerk of works (COW) or resident engineer (RE), using his copies both of the drawing and the schedule, will check that the reinforcement is correct in size, shape and position prior to pouring the concrete.

What are material lists used for?
Upon completion, the requisition lists are forwarded to the material ordering office where they are compared with the stock lists of available materials. This makes clear which steel sections, plates and bolts have to be ordered from a steel stockholder and which have to be ordered direct from the rolling mills.

Cutting lists are prepared by the detailer, forwarded to and completed in the material ordering office, and sent to the fabricating shop and the works office. The required material is taken from the stockyard, brought to the fabricating shop and the members and connections cut exactly as specified on the list. When the member has been drilled and the end connections fitted the total material in the assembled member must correspond to that shown on the requisition lists.

The weight of each individual assembled element can also be calculated from the information given on the material lists, and this is useful in determining methods of transport and erection as well as the final costing.

What is a bolt list used for?
These prepared lists are sent from the drawing office to the office and stores where the availability of bolts of various types is checked and the material ordered from a stockholder if necessary. A copy is also sent to the fabricating shop where the bolts are used for shop assembly connections, e.g. cleats to beams or gussets for trusses.

The structural steelwork is delivered to the site for erection together with bags of bolts corresponding to the details and quantities shown on the site bolt list. The erectors check whether the bolts delivered are in accordance with the list of site bolts and report any discrepancy to the office.

Again the cost can be readily calculated from the information given both for shop and site bolts.

FORMAT FOR DRAWINGS AND SCHEDULES

There was a time when almost every major structural consultant evolved his own method of detailing and scheduling structural information. As far as structural steelwork is concerned a number of methods exist which by tradition and use have become accepted as standards. However when dealing with reinforced concrete, a definite attempt has been made to standardise all detailing within the profession. This was done by a committee drawing its members from the Concrete Society and the Institution of Structural Engineers, and resulted in the publication of an extremely valuable report entitled the *Standard Method of Detailing Reinforced Concrete*. The recommendation of this report, and another by the Concrete Society entitled *Standard Reinforced Concrete Details*, are used throughout this book.

In the case of scheduling reinforcement, BS4466 defines a standard method of scheduling and gives a set of bar shapes which, when suitably combined, cater for all detailing situations. Guidance for maximum and minimum areas of steel, stop-off positions for reinforcement, suitable bond and anchorage lengths, spacing of bars and cover required may be found in CP110 *The Structural Use of Concrete*. These are summarised in Chapter Seven. For a summary of salient points affecting structural steelwork see Chapter Three.

WHY HAVE STANDARD METHODS?

There are four reasons why every structural detailer should use the same method:

(*a*) standardisation;
(*b*) simplicity;
(*c*) speed;
(*d*) saving.

These four reasons are in progressive order, i.e. the better the standardisation the greater the simplicity, which results in greater speed. The outcome of this should be greater efficiency and hence a saving in costs. This is true of both the design office and construction on site.

The advantages of a standard method of structural detailing to the contractor, the consultant and, of course, the detailer are as follows.

The contractor

The possibility of a contractor working with one consultant exclusively over a period of years is remote. Prior to standard methods being recognised the contractor had to re-educate his employees as they went from job to job, i.e. from consultant to consultant. It is evident that standardisation has been a great help to the contractor and his men, some of whom have difficulty in grasping the basis of one method, without the added problem of changing methods frequently.

The consultant

The system also has attractions for the consultant. Very often in a design office there will be more than one detailer working on any one job. Obviously great confusion would ensue if each was left to his own devices when producing working drawings, and the result would be costly to the consultant in terms of explanatory phone calls and site visits. It also means that the consultant can standardise his own drawing sheets and schedule paper. With the increased use and potential of computers in detailing, this standardisation becomes essential.

The structural detailer

Even the detailer can find an advantage in a standard method as he can move from one firm to another without having to learn a new method of detailing.

SELF-ASSESSMENT QUESTIONS

1. Discuss why it is necessary for the following people to have the structural engineer's drawings: (*a*) the quantity surveyor; (*b*) the services engineer; (*c*) the main contractor.

2. List seven people who use bar-bending schedules, describing how each uses his copy.

3. Describe the sequence of events which takes place from the time the drawing process starts to the construction on site of a structural steel frame.

4. What advantages are there in having standard methods for producing drawings?

CHAPTER TWO

How It's Done

<table>
<tr><td>CHAPTER OBJECTIVES</td></tr>
</table>

CHAPTER OBJECTIVES

After studying this chapter you should be able to:
* represent a structural detail in accordance with standard practice.

Size of drawings

To help in the clear communication of information only the recommended sizes of drawings should be used. These are as listed in Table 1, but note carefully that the A0 size should only be used in exceptional circumstances.

TABLE 1. RECOMMENDED SIZES OF DRAWINGS

British Standard reference	Size in mm
A0	841 × 1,189
A1	594 × 841
A2	420 × 594
A3	297 × 420
A4	210 × 297

Note the interrelationship of the sizes stated which allows for any standard size of drawing to be neatly folded to fit an A4 size envelope for posting. The fact that the drawings are all proportional means they are easily reduced and enlarged as required.

Within the overall sizes shown above, all

MATERIAL DIFFERENCES

When standard methods of detailing structures are under consideration it is essential to differentiate between reinforced concrete and structural steelwork. Reinforced concrete detailing is very well documented in the report published by the Cement and Concrete Association for the Concrete Society and the Institution of Structural Engineers and entitled *The Standard Method of Detailing Reinforced Concrete*. We will refer to this document merely as "the Report" and its recommendations are adopted throughout this book. Structural steelwork is not so well documented; however, there are methods which are recognised as standard within the industry.

Since BS1192:1969 *Drawing Office Practice for Architects and Builders* is the basis for some of the standards adopted in both disciplines, there are areas of similarity, but the two materials are so diverse there must be areas in which they are different. We shall look therefore at the areas of similarity first and later treat each material separately.

COMMON STANDARDS

The common areas are the size of drawings, scales, spelling, lettering, lines, line thicknesses, dimensions, sections and levels. These are documented in BS1192 and the Report and are, in the main, adopted by the structural steelwork industry. We shall deal with each of these in turn.

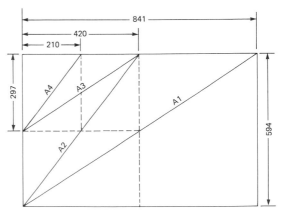

Fig. 5. *Interrelationship of drawing sizes* (all dimensions in mm).

drawings should have a 25 mm filing margin on the left-hand side and, with the exception of the A4 size, there should be a 10 mm margin on the other three sides. The A4 drawing does not require these 10 mm margins.

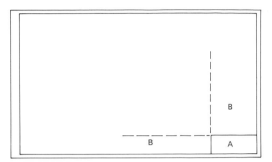

Fig. 6. *Position of information panel (A) on a drawing.*

An information panel is required on each drawing. This is shown as panel A in Fig. 6 and should be 180 mm × 60 mm, placed in the bottom right-hand corner of the drawing sheet. The following information should be included:

    (*a*) job title;
    (*b*) drawing title;
    (*c*) drawing number;
    (*d*) name of firm.

Additional information may be included in panel B (*see* Fig. 6) and could include such information as:

    (*a*) scales;
    (*b*) drawn by;
    (*c*) checked by;
    (*d*) date;
    (*e*) notes;
    (*f*) revisions.

Panel B is more usually developed vertically, but may be horizontal if this is more suitable.

Scales

The following scales are normally adopted:

    (*a*) site layout and simple general arrangement—1 : 200;
    (*b*) general arrangement—1 : 100 or 1 : 50;
    (*c*) simple wall and slab details—1 : 50;
    (*d*) beam and column elevations—1 : 20;
    (*e*) beam and column sections and details—1 : 20 or 1 : 10;

    (*f*) where larger scales are desirable—1 : 5, 1 : 2 or full size (FS).

Lettering and spelling

The detailer must always be working to produce clear and legible drawings. This is most important when we consider how a print of the original drawing is used, or abused, on site. After repeated unfolding and refolding, use by operatives with muddy or oily hands, being taken out in the wind and rain, these drawings can end up in a sorry state and yet they still have to be readable. The lettering used must therefore be simple and unaffected. Where the detailer's freehand lettering is not neat or easily readable stencils or transfers may be employed. The inexperienced detailer should draw two parallel lines at a constant distance apart to guide him in his lettering. The size of letters may be 5–8 mm for titles and 1.5–4 mm for notes. The Report recommends capital letters for titles and subtitles, with lower-case letters for notes. Notes look better when arranged in a panel together, e.g. panel B in Fig. 6, rather than scattered over the drawing.

The spelling of all words should be in accordance with *The Little Oxford Dictionary*, taking the first spelling where alternatives are given, e.g. asphalt, kerb, lintel, etc.

Lines and line thickness

When drawing in ink there are three pens which the detailer should possess, each yielding a different line thickness; these are 0.2 mm, 0.4 mm and 0.6 mm. The recommended thicknesses are:

    (*a*) main reinforcing bar—0.6 mm;
    (*b*) general arrangement drawings, outlines and profiles in section, primary elements in a steel drawing, e.g. load-bearing walls and structural slabs—0.6 or 0.4 mm;
    (*c*) outlines of components and general steel details, secondary elements in a steel drawing, e.g. partitions, concrete outlines in a reinforcement drawing, links and stirrups in a reinforcement drawing—0.4 mm;
    (*d*) dimension lines, grids, hatching, breaklines—0.2.

The cross-section of bars should be drawn approximately to scale.

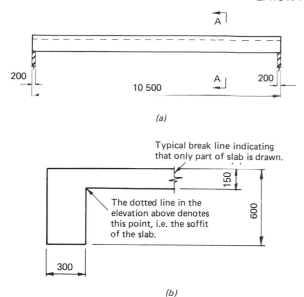

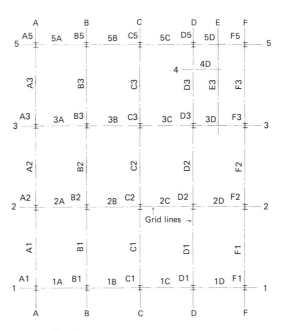

Fig. 7. *Use of dotted line to show hidden detail.* (*a*) Elevation of beam. (*b*) Section across AA.

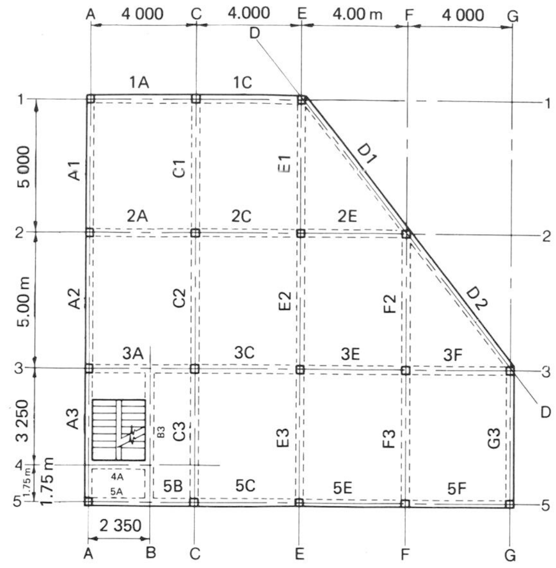

Fig. 8. *The use of grid lines.*

Hidden details
Hidden details are denoted by a broken line, as illustrated in Fig. 7.

Break lines
Where only part of an element is shown it is denoted as illustrated in Fig. 7.

Centre lines
These are denoted by a chain dotted line which usually projects a distance beyond the outline, allowing dimensions to be added in a position which does not obscure the profile of the structure. These are illustrated in Figs. 8 and 9 as grid lines.

Line density
Those drawings which are done in ink present little or no problem in reproduction. Occasionally in structural steelwork pencil drawings are produced, and even for microfilm reproduction leads not harder than 2H are satisfactory.

Dimensions
The dimensions ought to be placed so that they can be read when viewed from the bottom or the right-hand side of the drawing. The dimension lines should be continuous and not interrupted

Fig. 9. *A modified system of grid lines.*

for the insertion of the dimension figures. These figures are to be placed immediately above the corresponding dimension line, as near the centre as possible and arranged along the line, not at right angles to it.

For site layouts and simple general arrangements the metre may be used as the unit. However for small sections and detailing reinforcement or connections the millimetre is recommended. As far as possible values in millimetres should be expressed as multiples of 5 and it is not necessary to write mm. Various ways of showing dimensions are illustrated in Figs. 7, 8 and 9.

Sections

The directions of sections should be taken looking consistently in the same direction. This is usually looking left for beams and down for columns. There are, however, many situations where it is necessary to take a section looking from some other direction in order to illustrate what is required more clearly. This practice is acceptable but must be kept to a minimum.

The section should be drawn as near as possible to the detail to which it relates and be described by letters, rather than numbers, which ought to be readable with the drawing in its usual orientation. The section marks are shown in Fig. 7.

In reinforced concrete detailing it is usual to apply a light shading of a coloured pencil, normally green, to the back of the negative over the area of the section and this subsequently prints as a light grey area. This shading must not be applied too heavily otherwise the details within the section will be obscured or even obliterated.

Levels

Levels record the distance of a position above or below a defined datum. It is usually not necessary to relate a job datum to the Ordnance Survey datum unless the work is a major civil engineering or building project. On smaller works a suitable fixed point is taken as datum and all other levels related to it. Those levels should be expressed as multiples of 5 mm or 10 mm as appropriate. Figure 10 illustrates how these levels should be indicated, both on plan and section or elevation.

Finished floor levels or structural floor levels should be indicated as follows:

FFL 11.705 m   SFL 11.500 m

Existing levels should be indicated as follows:

EL 10.945 m

(a)

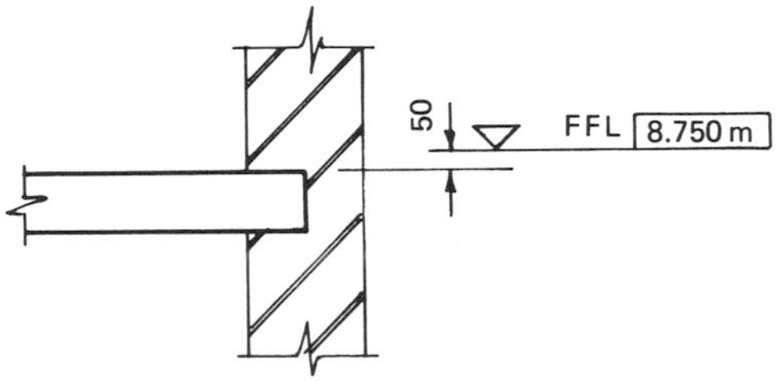

(b)

Fig. 10. *Indicating levels:* (a) on plan; (b) on section.

Representation of materials

There are two methods of representing materials, either graphically or by using abbreviations.

The relevant graphical symbols are shown in Fig. 11, while the abbreviations are listed in Table 2.

TABLE 2. ABBREVIATIONS FOR MATERIALS

Material	Abbreviation
Asbestos	abs
Asbestos cement	abs ct
Asphalt	asph
Brickwork	bwk
Building	bldg
Column	col
Concrete	conc
Drawing	drg
Foundation	fdn

Letters and signs

Many abbreviations and symbols are documented in British Standards such as BS499, *Welding Terms and Symbols*, and BS1991, *Letters, Symbols, Signs and Abbreviations*. Some of the more commonly used symbols in relation to structural detailing are shown in Fig. 12.

Having considered the common ground

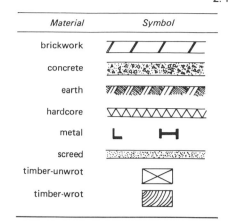

Material	Symbol
brickwork	
concrete	
earth	
hardcore	
metal	
screed	
timber-unwrot	
timber-wrot	

Fig. 11. *Graphical symbols used to represent materials.*

Description	Symbol
Centre to centre	C/C
Centre line	₵
North point	
Rise of stair	1 2 3 4 5 6 7 8
Finished floor level	FFL
Ground level	GL
Approximate	approx
Diameter	dia
Inside diameter	I/D
Outside diameter	O/D

Fig. 12. *Some of the more commonly used symbols used in structural detailing.*

between reinforced concrete and structural steelwork detailing we must now treat each individually.

REINFORCED CONCRETE DETAILING

As stated above, the reinforced concrete detailing shown in this book complies with the Report. It is therefore essential that the principles outlined are clearly grasped. We may profitably consider the scope of the Report, its standards, symbols and scheduling.

The scope of the Report

The Report covers ways in which detailing can be standardised and simplified, basically for reinforced concrete, although the principles outlined may be applied to prestressed concrete. It is appreciated that many design principles arc grasped by the competent and experienced detailer but these are not covered by the Report. Some of the basic principles are discussed further in Chapter Seven.

The standards of the Report

A lot of the standards recommended have been outlined already in the previous section of this chapter. There are, however, two which have a particular application to reinforced concrete.

Types of drawing

There are two broad categories of drawing which are suggested, viz. general arrangement drawings and detail drawings. The former consist of plans, elevations and sections showing the setting out, layout, dimensions and relationship of all members. The latter requires more explanation.

On site there are principally two tradesmen who use the structural detail drawings: the joiner, who is involved with the shuttering of formwork; and the steelfixer, who positions and ties the reinforcement. The joiner is most interested in the concrete profile, including the positions and sizes of any holes and fixings, whereas the steelfixer is more interested in information relating to the location of the reinforcement. To avoid confusion and ambiguity in the more complex drawings it is recommended that outline drawings are kept separate from the reinforcement drawings. This can easily be accomplished by using the following method.

The detailer prepares the outline drawing in ink with the dimensions in soft pencil. At this stage a copy negative is produced which gives the impression that the complete drawing has been prepared in ink. This becomes the finished article for the joiner's purposes and the normal paper prints may be made. The pencil dimensions are then rubbed off the original negative and the reinforcement details put on in ink. Thus we have two finished drawings, each capable of reproduction, without a significant amount of extra drafting work.

Grid lines

A reference system for the structural fabric of a building is most desirable since it removes a large area of possible ambiguity and error. The most common system used both for reinforced concrete and structural steelwork is that employing grid lines. For reinforced concrete the Report recommends one method but there are others, some of which are used for structural steelwork. Here we are considering the method used within the Report; another is dealt with later in this chapter (*see* Fig. 9).

Grid lines are chain dotted lines drawn at, or close to, lines of beams or centres of columns. The system is flexible enough to take account of non-standard cases or skewed buildings, which means that the existence of a grid line does not necessarily imply that there is a beam or a column on that line.

Grid lines are usually numbered in one direction and lettered in the other, the top left-hand corner being a convenient starting point, as shown in Fig. 8.

The columns are numbered thus: A1 at the intersection of grid line A and grid line 1; F2 at the intersection of grid line F and grid line 2, etc.

The beams can simply be referred to as "beam on grid line 3", etc. If more detail is required, as in precast concrete or structural steelwork, then each bay may be distinguished as 3A, 3B, 3C, etc., or A1, A2, A3, etc. Note that capital letters are used for reference. Formerly lower case letters were employed but these are not acceptable for use in computing systems.

A further letter may be incorporated to give a floor reference. It is usual to begin at the ground floor with the letter A and move upward (B, C, D, etc.). If there are levels below the ground floor, e.g. a basement, these are referred to as 1, 2, 3, etc. Thus the beam on grid line 2 between grid lines C and E at second floor level is referred to as C–2C.

The symbols of the Report

The Report recommends the abbreviations shown in Table 3, but also notes that if there is a possibility of confusion or ambiguity then the words should be written in full. If any other abbreviation is to be used it must be clearly defined on the drawing. Some of those shown

have been described earlier when dealing with the symbols which are applicable to both reinforced concrete and structural steelwork. Those now listed are the only ones recognised by the Report.

TABLE 3. ABBREVIATIONS RECOMMENDED BY THE REPORT

Term	Abbreviation
General abbreviations:	
reinforced concrete	RC
brickwork	bwk
drawing	drg
full size	FS
not to scale	NTS
diameter	dia
finished floor level	FFL
structural floor level	SFL
existing level	EL
horizontal	hor
vertical	vert
Abbreviations relating to reinforcement:	
each way	EW
each face	EF
far face	FF
near face	NF
bottom	B
top	T

An important recommendation of the Report is that of notation. When referring to reinforcement on drawings the sequence of description should be: number, type, size, mark, centres, location or comment. To describe twenty-five high-yield high-bond bars of sixteen millimetre diameter at two hundred millimetre centres on each face of a wall, the recommended abbreviation is 25T16–31–200 EF.

The number 31 is called the bar mark and is used to give each bar a unique reference. It must only be used for bars which are identical in every respect, i.e. same diameter, type of reinforcement, length, shape and bending. There may be quite a few bar marks numbered 31 on different drawings, but when these numbers are combined with the bar schedule reference number then bar mark number 31 on drawing 1 is distinguished from bar mark number 31 on drawing 2. On small jobs where there are only a few drawings it may be convenient simply to start at bar mark 1 and carry on through the whole job in consecutive sequence.

The symbols for the type of reinforcement should be as follows and, if necessary, more fully described in the specification:

R—plain or deformed grade 250 bars complying with the requirements of BS4449;

T—type 2 deformed bars of grade 460/425, complying with the requirements of BS4449 or BS4461;

X—a general abbreviation for types not covered by R or T.

An explanation of X is required not only in the specification but also on the drawings and schedules.

Schedules in the Report

The reasons for bar schedules have been outlined in general in Chapter One. There are, however, a number of important points which must be understood to produce a satisfactory schedule.

Bar sizes and lengths

The preferred diameters of bars, expressed in millimetres, are 6, 8, 10, 12, 16, 20, 25, 32 and 40 mm. Reinforcing bars are available in lengths up to 18 m, but over 12 m is only obtained by special arrangement with the supplier. However, due to transportation difficulties, lengths over 10 m are used only in exceptional circumstances. Also to facilitate transportation each bent bar should be contained in an imaginary rectangle, the shorter side of which should not be greater than 2,750 mm. It is convenient to use lengths of 8 m or 9 m for bars of larger diameter, with 6 m as a maximum for bars less than 10 mm in diameter. If the bars are to be cut from stock lengths consideration must be given to the wastage. If 8 m-long bars are being cut from 12 m stock lengths there will obviously be 4 m lost. If these lengths are not required elsewhere on the job the competent detailer would adjust his details to minimise the wastage. Similarly, with short cut lengths for links, for example, a small adjustment to the dimensions can mean that a stock length of bar can be cut to give an exact number of lengths.

Where reinforcement is prefabricated it is important to have the cages as stiff as possible

to facilitate lifting and transportation. To achieve this it is helpful to use the largest possible bar size.

Often in floor slabs or walls which are lightly loaded or in ground floor slabs the reinforcement required is in the form of a standard factory produced fabric sheet. BS4466:1981 lists three categories of fabric, viz.:

(*a*) designated fabric;
(*b*) scheduled fabric;
(*c*) detailed fabric.

The first of these, the designated fabric, is perhaps the most common and is identified by a standard fabric reference, e.g. A193, B385, C503, etc. (*see* Table 21, page 87). The scheduled fabric is defined as fabric reinforcement with a regular wire and mesh arrangement that can be defined by specifying the size and spacing in each direction. BS4466:1981, Fig. 1, shows the method for doing this. A detailed fabric is defined as fabric reinforcement not covered by the other two categories. A dimensioned drawing is required to define the mesh arrangement.

When standard factory produced fabric sheet is being scheduled it should be done on special sheets, exclusively by itself, and a sketch given in the column headed *Bending instruction* to ensure that the fabric is bent in the proper manner. It is also recommended that the fabric be grouped according to the size of sheet and weight per square metre.

Where fabric is used as the reinforcement the Report recommends the following method of detailing.

The outline of each sheet is drawn as a rectangle, with a diagonal line to indicate its extent. The location of the sheet, whether top or bottom, is shown by the direction of the diagonal. Where this could cause confusion separate plans showing top or bottom reinforcement may be drawn, or the suffix T or B added to the sheet mark.

The reference mark of the sheet is written along the diagonal and a double-headed arrow indicates the direction of the main wires.

Where two layers of fabric are used in conjunction with each other in the top or bottom of a slab the layers should be nested.

In section the fabric is shown by heavy dotted lines. Where the section is parallel to the

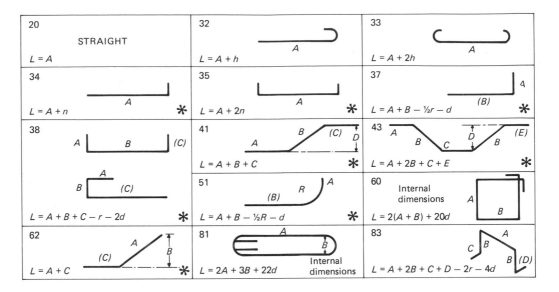

(a)

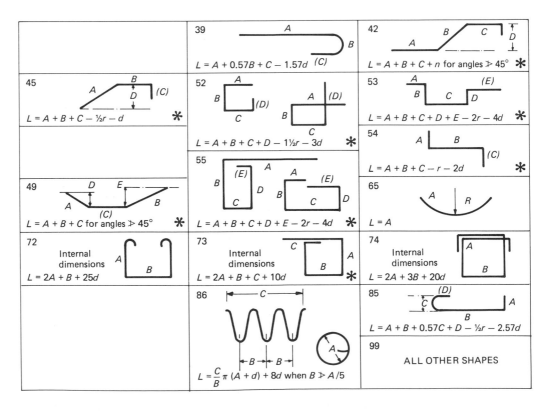

(b)

Fig. 13 (*a*) *Preferred shapes.* (*b*) *Other shapes.* Bracketed dimensions are run-out dimensions and may be omitted from the final schedule. For shape codes 60, 72–5 and 81 internal dimensions are assumed; for all other shapes outside dimensions are assumed unless otherwise marked. *For these shapes in particular the effect of a positive cutting tolerance increasing the actual length of the "free" leg or legs by up to 25 mm should be considered.

main wires, long dashes are used, whereas short dashes indicate the section is taken at right angles to the main wires.

Bar shapes

In the past the detailer often had to draw a dimensioned sketch of the bar he required. With computers becoming more common in detailing a list of preferred shapes has been drawn up and each of these given a number. This information is contained in BS4466 but the common shapes are illustrated in Fig. 13.

making allowances for bends and hooks (*see* Table 4). The total length is calculated according to the formula given, rounded to the nearest 25 mm, and entered in the schedule sheet under the column headed *Total length*.

Bends and laps in reinforcement

The cost of bent reinforcement is greater than that of straight bars. It is therefore more economic to detail the reinforcement with as few bends as possible, thus providing relatively simple details.

TABLE 4. MINIMUM FORMER RADII, BEND AND HOOK ALLOWANCE

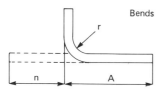
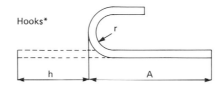

Bar size	Grade 250 bars complying with the requirements of BS4449 (type and grade R)			Grade 460/425 bars complying with the requirements of BS4449 or BS4461 (type and grade T)			Hard drawn wire complying with the requirements of BS4482			
d	r†	n	h	r†	n	h	d	r†	n	h
6‡	12	100	100	18	100	100	5	15	100	100
8	16	100	100	24	100	100	6	18	100	100
10	20	100	100	30	100	110	7	21	100	100
12	24	100	110	36	100	140	8	24	100	100
16	32	100	150	48	100	180	9	27	120	135
20	40	100	180	60	110	220	10	30	120	135
25	50	130	230	100	180	350	12	36	130	145
32	64	160	290	128	230	450	—	—	—	—
40	80	200	360	160	280	560	—	—	—	—
50‡	100	250	450	200	350	700	—	—	—	—

* The use of hooks as end anchorages is not normally necessary with deformed bars.

† It should be borne in mind that special precautions such as reducing the speed of bending or increasing the radius of bending may be necessary in cold weather. For bending and rebending on site refer to CP110.

‡ Non-preferred size.

NOTE. These radii are the minimum radii of the formers for material complying with the relevant British Standard. Certain design situations may call for larger radii and for these, the recommendations in BS5400, Part 8, and CP110 should be followed.

When preparing a bar schedule the detailer chooses the shape corresponding to the one he requires and enters the number in the column headed *Shape code*. The columns headed *A*, *B*, *C*, etc., on the schedule sheet refer to the dimensions *A*, *B*, *C*, etc., on the sheet of preferred shapes. These are given to the nearest 5 mm,

Where a section is heavily reinforced special care must be taken at lap positions to ensure that it is possible to place and compact the concrete. It may be necessary to stagger the laps.

Tolerances

An important and yet sometimes much neglected

topic when considering scheduling is that of tolerances.

The cover to the reinforcement is liable to variation due to small errors in the positioning of shuttering and in the cutting, bending and fixing of the steel. The cumulative effect of these errors can be very serious if an allowance is not made for this effect by giving tolerances. Table 5 shows the deductions necessary, both on the reinforcement and the member.

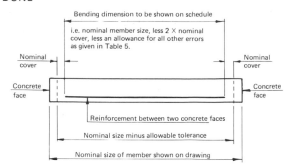

TABLE 5. DEDUCTIONS NECESSARY TO ALLOW FOR ERRORS IN THE COVER TO THE REINFORCEMENT

Overall concrete dimension in the direction of the tolerance (mm)	Deduction to determine the bending dimension (mm)	Tolerance on member size (mm)
Bent bars:		
0–1,000	10	not more than 5
1,000–2,000	15	not more than 5
over 2,000	20	not more than 10
Ends of straight bars:		
any length	40	not more than 10

Where a reinforcing bar is required to fit between two concrete faces, the dimension shown on the bar schedule should be determined as the nominal dimension of the concrete member, less the nominal cover at each side, less an allowance for all other errors in accordance with Table 5. If the tolerance on the member size exceeds the values shown, then either the deduction or the nominal cover should be increased. Figure 14 gives examples of how the lengths for the schedules should be calculated.

Reference numbers
In the past schedule sheets have sometimes been numbered by page number, sheet number, etc., and, if not properly cross-referenced to the relevant drawing, this has led to confusion. It is now recommended that they should have simple consecutive reference numbers not exceeding six characters and should be cross-referenced to the relevant drawing number. The first three characters refer to the drawing number, the next two to the schedule number and the last can be used for revision letters. For example the number 14307C appearing on a schedule refers to drawing number 143, schedule sheet 07, revision C. When this is incorporated with the bar mark it can be appreciated that every bar has a unique and distinct reference mark.

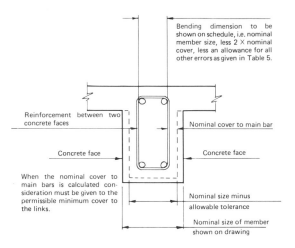

Fig. 14. *Calculating the dimensions of members to allow for cover and for errors.*

When it is necessary to revise a bar dimension on the schedule, the incorrect figures should be crossed out but not erased, and the correct figure written in above the wrong figure. The last letter in the schedule reference should change and this same letter written in the right-hand margin of the schedule in line with the corrected dimensions. This is illustrated in Fig. 15.

CURPE CONSULTANTS

Bar Schedule ref. | 0 | 1 | 5 | | 0 | 2 | Rev. | A |

Site Ref. **DUNFIELD SCHOOL**   Date:

Member	Bar mark	Type & size	No. of mbrs.	No. in each	Total No.	Length of each bar mm*	Shape code	A mm+	B mm+	C mm+	D mm+	E/r mm+
A1, B1.	1	T25	2	4	8	4250	20	4250				
	2	T25	2	6	12	3525	20	3525				
	3	T16	2	3	6	3775	20	3775				
	4	T16	2	3	6	2125	37	300				
	5	T25	2	4	8	3600	51	450				150
	6	R10	2	28	56	2000	60	600	300			
	7	R10	2	14	28	~~1800~~ 1700	60	600	~~200~~ 150				A
	8	T20	2	6	12	2500	20	2500				
	9	T25	2	4	8	2950	41	1000	400		55	
	10	T20	2	6	12	3225	41	800	300		45	
	11	T25	2	6	12	5575	43	1000	400	2765	55	

All bending dimensions are in accordance with BS 4466

+ Specified to the nearest 5 mm
* Specified to the nearest 25 mm

Fig. 15. *How to revise a bar dimension on a schedule.*

STRUCTURAL STEELWORK DETAILING

While not documented to the same extent as reinforced concrete detailing, methods of detailing structural steelwork have evolved within individual firms in the industry and are now, by tradition, considered as standard, being recognised by the personnel in the drawing office, general and works offices, template shop or loft, fabricating shop and by the erection staff.

Prior to the structural detailer becoming involved, a design drawing showing the sizes of the various members should have been completed by the structural designer. The detailer takes that drawing and prepares the shop detail drawings which give detailed instructions for the fabrication of all the members and connections.

The standards set out under the heading *Common Standards* in this chapter are utilised, plus the relevant portions from this section.

Grid lines

As stated when dealing with reinforced concrete detailing, there are great advantages in having a simple grid system which enables each member to be labelled unambiguously. The method illustrated in Fig. 8 can also be used for structural steelwork, or a modified version as shown in Fig. 9 can be employed.

Representation of structural steel sections

Structural steel sections are normally represented as shown in Fig. 16.

Section	Abbreviation	Pro-file	Example of designation
Universal beam	UB	I	254 × 146 × 31 UB
Universal column	UC	I	203 × 203 × 46 UC
Rolled steel joist	RSJ, JOIST	I	178 × 102 × 21.54 Joist
Rolled steel channel	RSC	[	178 × 76 × 20.84 [
Rolled steel angle	RSA	L	70 × 70 × 6 L
Rolled tee	Tee	T	51 × 51 × 4.76 T
Structural tee	Tee	T	127 × 152 × 19 Struct T
Steel plate	PLT	—	200 × 100 × 10 PLT
Circular hollow section	CHS	O	60.3 × 5 CHS
Rectangular hollow section	RHS	□	150 × 100 × 5 RHS
Steel flat	flat	—	200 × 10 flat

Fig. 16. *Representation of structural steel sections.*

Structural steelwork connections

There are three methods of connecting structural steelwork elements, viz.:

(*a*) bolting;
(*b*) riveting; and
(*c*) welding.

To facilitate these methods symbols for each have been devised (*see* Figs. 17 and 18). Although

Material	Elevation	Section
Site bolts, open holes		
Site bolts, open holes, CSK NS		
Site bolts, open holes, CSK FS		
Shop bolts		
Shop bolts CSK NS		
Shop bolts CSK FS		
Rivets		
Rivets CSK NS		
Rivets CSK FS		
High strength friction grip shop bolts		
High strength friction grip site bolts		

Fig. 17. *Symbols for bolts and rivets.* CSK NS indicates countersunk near side; CSK FS indicates countersunk far side.

Designation	Illustration	Symbol
Butt weld between flanged plates (the flanges being melted down completely)		
Square butt weld		||
Single-V butt weld		V
Single-bevel butt weld		
Single-V butt weld with broad root face		Y
Single-bevel butt weld with broad root face		
Single-U butt weld		
Single-J butt weld		
Backing or sealing run		
Fillet weld		
Flat (flush) single-V butt weld		
Convex double-V butt weld		
Concave fillet weld		
Flat (flush) single-V butt weld with flat (flush) backing run		

Fig. 18. *Diagrammatic representation of various types of weld.*

rivets are now practically obsolete in structural steelwork for buildings they are still included in standard handbooks and so they are shown in Fig. 17.

Bolts
Three types of bolts are encountered by the detailer of structural steelwork, black bolts, close tolerance and high strength friction grip bolts. Each may be further sub-divided depending on use, i.e. shop bolts for shop connections and site bolts for the use by the erector on site.

Welding
It is not intended to deal fully with the technicalities of welding. The information presented is that which will be of use to the structural detailer.

In general terms there are two basic types of weld:

(a) the fillet weld; and
(b) the butt weld.

The former is approximately triangular in cross section while the latter may be square, V or U shaped.

The type of weld may be indicated on the drawing by any one of several methods. The most usual ways are as shown.

(a) Use of symbols from BS499 in conjunction

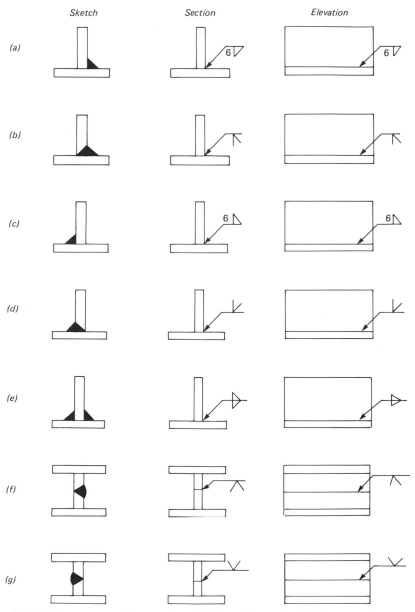

Fig. 19. *The use of welding symbols.* (*a*) 6 mm fillet weld. (*b*) Single bevel butt weld. (*c*) 6 mm fillet weld. (*d*) Single bevel butt weld. (*e*) Fillet welds. (*f*) Single V butt weld. (*g*) Single V butt weld.

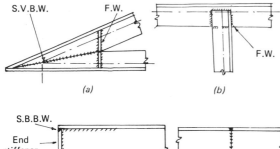

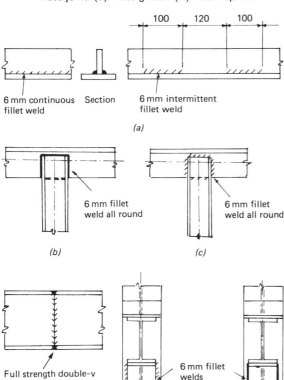

Fig. 20. *Use of abbreviations for welds.* (a) Truss shoe. (b) Truss joint. (c) Plate girder. (d) Beam splice.

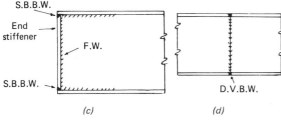

with a reference line and arrow. In this British Standard a comprehensive list may be obtained; however, some examples are shown in Fig. 18. Figure 19 illustrates the use of the symbols in Fig. 18 together with the reference line and arrow. Further examples are given in BS499.

(b) Abbreviation of the name, together with thick lines or short lines at 45° slope. This method utilises the letter symbols shown in Table 6, and illustrated in Fig. 20.

TABLE 6. ABBREVIATIONS FOR DIFFERENT TYPES OF WELD

Type of weld	Abbreviation
Fillet	FW
Square butt	SB
Single V butt	SVBW
Double V butt	DVBW
Single U butt	SUBW
Double U butt	DUBW
Single bevel butt	SBBW
Double bevel butt	DBBW

(c) Full description of the weld. This is illustrated in Fig. 21.

Joint forms

Each of the above types of weld can be used with a number of forms of joint. These are detailed in Fig. 22.

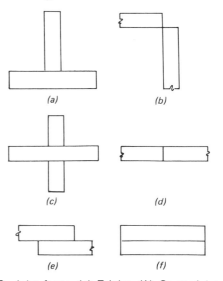

Fig. 21. *Use of full description of weld.* (a) Fabricated T section. (b) Truss joint. (c) Truss joint (alternative). (d) Beam splice. (e) Beam to stanchion connection (alternatives).

Fig. 22. *Joint forms.* (a) T joint. (b) Corner joint. (c) Cruciform joint. (d) Butt joint. (e) Lap joint. (f) Edge joint.

SUMMARY

We have seen why drawings are necessary and how various parts of a drawing should be represented both for reinforced concrete and structural steelwork. We will now treat each material separately and show in detail how these principles are applied in detail to various structural elements.

SELF-ASSESSMENT QUESTIONS

1. Show how an A1 size drawing sheet could be folded to fit an A4 size envelope, with the drawing title on the outside.

2. A reinforced concrete beam is 500 mm deep by 250 mm wide and spans 8.750 m between the centres of supporting columns which are 250 mm square. A slab which is 150 mm thick cantilevers from one side of the beam a distance of 1.000 m from the edge of the beam. Draw an elevation and section of the beam showing all dimensions and sizes utilising the requirements as set out in the preceding chapter.

3. Sketch how the following materials are indicated on a drawing: (*a*) brickwork; (*b*) concrete; (*c*) earth; (*d*) hardcore; (*e*) screed; (*f*) wrot timber.

4. (*a*) A straight reinforcing bar is required to fit between two concrete faces which are 7.250 m apart. If the nominal cover to the reinforcement is 40 mm, calculate the length of bar to be supplied. (*b*) Steel links are required for a reinforced concrete column which is 450 mm × 300 mm in cross section. If the nominal cover to the link is 20 mm, calculate the external bending dimensions of the link.

5. (*a*) State three methods of indicating types of welds on a detail drawing. (*b*) Show the appropriate symbol from BS499 for the following welds: (*i*) fillet; (*ii*) single V butt; (*iii*) double V butt; (*iv*) double bevel butt. (*c*) Show the appropriate symbol for the following bolts: (*i*) site bolt; (*ii*) shop bolt; (*iii*) high strength friction grip; (*iv*) site bolt CSK FS.

6. What are the principal types of bolts used for structural steelwork connections?

CHAPTER THREE

Structural Steelwork in General

CHAPTER OBJECTIVES

After studying this chapter you should be able to:
* appreciate the grades and range of steel sections available;
* have sufficient knowledge of that part of workmanship, fabrication and assembly procedure which influences the preparation of shop detail drawings.

INTRODUCTION

Chapters Three to Six will be concerned solely with structural steelwork.

The sub-contractor responsible for the fabrication of the structural steelwork obtains his material from the rolling mills or the steel stockholder. In addition he normally carries his own stock of the more frequently used sections.

The structural steel sections are delivered to the fabricator's stockyard where they are checked against the appropriate order and then placed in a convenient position in the yard until required for fabrication.

After fabrication, assembly and painting are completed, the members and pieces are despatched to the site for the subsequent erection of the structural frame on prepared foundations.

This chapter is concerned with the range and selection of materials and the workmanship involved in the fabrication of members and connections. The information is of importance to the structural steel detailer in the preparation of detail drawings.

GRADES OF STEEL

The three grades of steel available are 43, 50 and 55. Grade 43 is mild steel and is the most commonly used for structures. Grades 50 and 55 are high yield steels and possess greater strength, but are more costly.

STANDARD STEEL SECTIONS

Hot rolled sections

The material is produced at a steel rolling mill where it is formed into a variety of shapes. The shape, dimensions and the weight of sections are standardised and their dimensions and properties are listed in structural steelwork handbooks. Figure 23 shows a range of typical structural steel sections.

Universal beams and columns

The size specified for any universal section is the serial size. Universal sections are rolled in several masses per metre length by varying the web and flange thicknesses but with the depth between the fillets remaining constant within the serial size. The overall size increases accordingly as the mass per metre length of the section increases. Both are designated by serial size × mass per metre length. Universal beams range from serial size 914 × 419 down to 203 × 133 and universal columns from 356 × 406 down to 152 × 152.

Structural tees are cut from universal beams and columns.

Joists are designated by nominal size × mass per metre length and range from 203 × 102 down to 102 × 64 with 5° taper flanges. Sections with 8° taper flanges are also available.

Channels are designated by nominal size × mass per metre length and range from 432 × 102 down to 76 × 38.

Equal and unequal angles are designated by leg length × leg length × thickness and range from 200 × 200 × 24 down to 25 × 25 × 3, and 200 × 150 × 18 down to 65 × 50 × 5 respectively. Full dimensions are included at the end of this chapter.

Hollow sections are available in circular, square and rectangular shapes and are designated by outside dimensions × thickness.

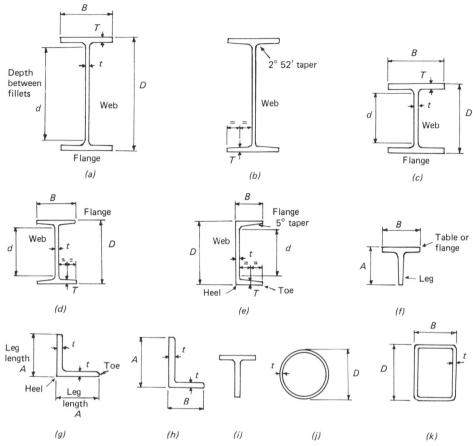

Fig. 23. *Typical structural steel sections.* (*a*) Universal beams, parallel flanges. (*b*) Universal beams, taper flanges. (*c*) Universal columns. (*d*) Joists. (*e*) Channels. (*f*) Rolled tees. (*g*) Equal angles. (*h*) Unequal angles. (*i*) Structural tees. (*j*) Circular hollow sections. (*k*) Rectangular hollow sections.

Castellated sections are fabricated from universal beams, universal columns, and joists and are shown in Fig. 24.

Compound sections
These are formed from standard structural sections connected together by means of welded, battened and latticed methods of construction. A few typical compound sections are shown in Fig. 25.

Cold formed sections
A range of cold formed sections are also available and are extensively used for roof purlins and sheeting rails where advantage is taken of their relative lightness in comparison with hot rolled

angles, channels and joists. Figure 26 shows typical examples.

TOLERANCES

Rolling tolerances
Continued use of the rolls at the steel mills result in wear and tear which may cause minor deviations from the specified standard dimensions and weights of sections. Permitted tolerances on dimensions for universal beams and columns range from $\pm$ 3.2 mm on depth and + 6.4 mm or − 4.8 mm on flange width. For specified depths of joists and channels, 3.2–4.8 mm over and 0.8–1.6 mm under is permitted. For a

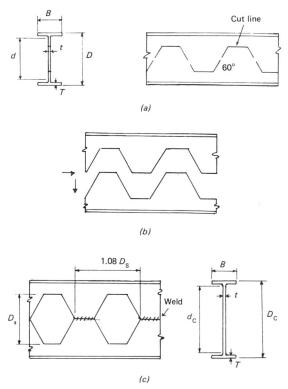

(a)

(b)

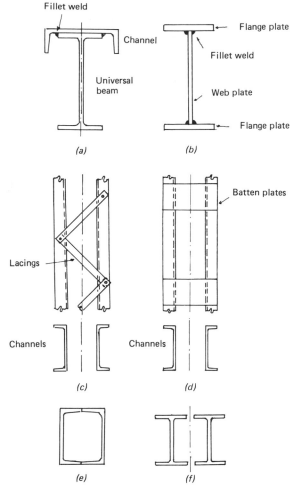

(a)   *(b)*

(c)   *(d)*

(e)   *(f)*

Fig. 24. *Typical castellated universal beam.* (*a*) Original section cut. (*b*) Separate and line up. (*c*) Final section.

$$D_C = D + \frac{D_S}{2}$$

where D_S = serial depth of original section.

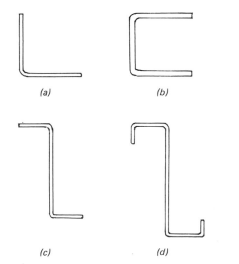

(a)   *(b)*

(c)   *(d)*

Fig. 26. *Typical cold formed sections.* (*a*) Angle. (*b*) Plain channel. (*c*) Zed section. (*d*) Lipped zed section.

Fig. 25. *Typical compound sections.* (*a*) Gantry girder. (*b*) Plate girder, autofab beam. (*c*) Latticed strut. (*d*) Battened strut. (*e*) Channels welded toe to toe. (*f*) Universal beams latticed.

circular hollow section the permitted tolerance on the outside diameter is $\pm$ 0.5 mm or $\pm$ 1 per cent, whichever is the greater. For a rectangular hollow section the permitted tolerance on the outside dimensions is $\pm$ 0.5 mm or $\pm$ 1 per cent, whichever is the greater.

It is necessary when marking off the position of holes and connections to allow for these possible variations from the specified dimensions.

Erection clearances

To facilitate the erection of steel beams to stanchions or to other beams a clearance of no more than 2 mm at each end should be given for:

(*a*) cleated ends of members;
(*b*) plated ends of members.

FABRICATION

Information regarding the cutting, notching, marking, holing, machining and painting of structural steelwork, with reference to detailing, will now be considered.

Cutting

From the stockyard the requisitioned materials are brought to the fabrication shop. The members, cleats, plates, gussets, etc., are then cut to lengths specified on the material cutting lists, or to actual template lengths and shapes where applicable.

Notching

In addition to cutting a member to its specified length, the end of an incoming or supported beam may require notching at its connection to a supporting beam. In the case of a beam to stanchion connection only the flanges may require notching. Figure 27 shows typical notches. A full range of notch and end clearance dimensions are given in structural steelwork handbooks. Examples taken from *Metric Practice for Structural Steelwork*, published by the British Constructional Steelwork Association (BCSA), are included at the end of this chapter (*see* Tables 12–16).

Marking

In order to provide a systematic procedure for detailing, fabrication and erection, each member

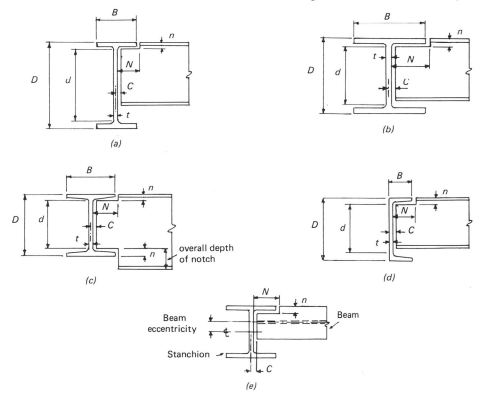

Fig. 27. *Typical notches.* (a) Universal beams. N is based upon the outstand from web face to flange edge, +10 mm to nearest 2 mm above, allowing for rolling tolerance. $n = (D - d)/2$, to nearest 2 mm above. $c = (t/2) + 2$, to nearest mm. (b) Universal columns. N, n and c are as for (a). (c) Joists. N = the outstand from web face to flange edge, +6 mm to nearest 2 mm above. $n = (D - d)/2$, to nearest 2 mm above. $c = (t/2) + 2$, to nearest mm. (d) Channels. N = outstand from web face to flange edge, +6 mm to nearest 2 mm above. $n = (D - d)/2$, to nearest 2 mm above. $c = t + 2$, to nearest mm. (e) Plan at beam connection to stanchion showing notching to top and bottom flanges of beam.

of the structure is given an identification or erection mark, e.g. B12, on the general layout and framing plans. This mark is also shown on the member on the detail drawings, painted on the member in the shop, and used by the steel erector on the site to enable the member to be properly positioned in the structure. For the purpose and ease of assembling each member and its connections together in the shop after fabrication is completed, an assembly mark, e.g. A, is given to each cleat, plate, gusset, cutting, etc., and shown on the detail drawings.

Holing

Terms used in the detailing of structural steelwork
It is important at this stage to define some terms which are commonly used in detailing (*see also* Fig. 28).

Pitch is the distance between adjacent bolts in a line parallel to the direction of stress in a member, and may be straight line or staggered pitch.

Gauge is the perpendicular distance between two consecutive lines of bolts parallel to the direction of stress. This distance is termed cross centres in the case of the flange of an I section.

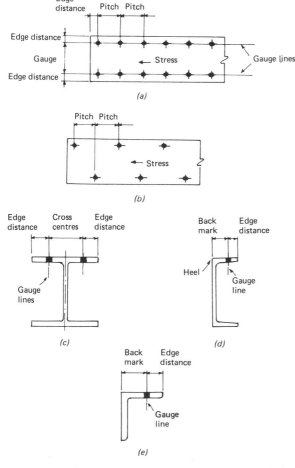

Fig. 28. *Some terms used in holing.* (a) Straight line pitch. (b) Staggered pitch. (c) I section. (d) Channel. (e) Angle.

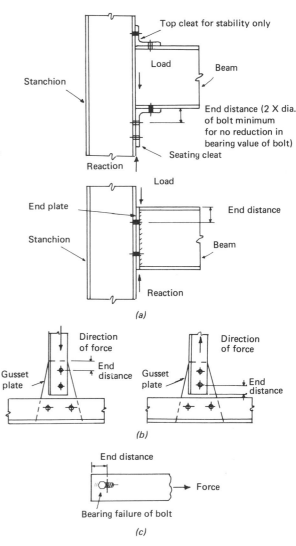

Fig. 29. *Examples of end distances.* (a) Beam to stanchion connections. (b) Connections for roof truss members. (c) Example of bearing failure of bolt.

Gauge lines are the lines along which bolts should be placed in the flanges of beams, columns, joists, channels, tees, and in the legs of angles.

Back mark (BM) is the perpendicular distance measured from the heel to the centre line (gauge line) of holes in the flange of a channel or in the leg of an angle.

Edge distance is the perpendicular distance measured from the centre line of a hole to the edge of a plate or section.

End distance is the edge distance in the direction in which a bolt bears. Examples are shown in Fig. 29.

Size of holes
Holes for black or high strength friction grip bolts should be 2 mm larger in diameter than the diameter of the bolt, up to 24 mm diameter bolts, and 3 mm larger for bolts over 24 mm diameter.

Holes for close tolerance bolts should be drilled to a diameter equal to the normal diameter of the shank, subject to a tolerance of $+0.15$ mm and -0 mm.

Spacing and edge distance
BS449 specifies rules regarding spacing and edge distance for holes, as follows.

Rules for spacing of holes. The minimum distance, centre to centre, of any two adjacent bolts should be not less than two and a half times the diameter of the bolt. Thin plates bolted to steel sections may buckle and distort when stressed if the bolts are too widely spaced. To prevent this occurring, maximum distances between bolts are specified, depending on the direction of stress in the member. (In all these cases t is the thickness of the thinner outside plate.)

(a) With bolts in a line adjacent and parallel to an edge of an outside plate, the smaller of 200 mm or 100 mm + 4t.

(b) With bolts in a line lying in the direction of stress: for tension members the smaller of 200 mm or 16t; and for compression members the smaller of 200 mm or 12t.

(c) Generally, the smaller of 300 mm or 32t.

It is important, for example when detailing welded end plate type of connections, to check the spacing of the connecting bolts accordingly.

If required, more bolts should be included than are necessary in order to reduce the spacing, or alternatively, thicker plates may be selected.

Rules for edge distance
Minimum edge distances are shown in Table 7, taken from BS449, Table 21. A smaller edge distance is required if the edge is rolled, machine flame cut, sawn, or planed, than if the edge is sheared or hand flame cut. For edges other than rolled edges it is suggested that the larger value is adopted as the method of cutting adopted is generally unknown or uncertain at the detailing stage.

To prevent distortion of plates the maximum edge distance is specified as 40 mm + 4t, where t is the thickness in mm of the thinner outside plate.

TABLE 7. MINIMUM EDGE DISTANCES

Diameter of hole (mm)	Distance to sheared or hand flame cut edge (mm)	Distance to rolled, machine flame cut, sawn or planed edge (mm)
30	50	44
24	38	32
22	34	30
20	30	28
18	28	26
16	26	24
14	24	22

Effect of end distance
If the end distance is less than the limit of twice the diameter of the bolt the allowable bearing stress for the bolt on the connected part should be reduced thus:

$$\text{Allowable bearing stress} \times \frac{\text{Actual edge distance}}{2 \times \text{Diameter of bolt}}$$

For example, if the end distance for a 20 mm diameter black bolt is 35 mm, then an allowable bearing stress of 250 is reduced to:

$$\frac{250 \times 35}{2 \times 20} = 218 \text{ N/mm}^2$$

Consequently the allowable load in bearing is reduced and this could reduce the allowable load for a bolt if bearing is the criterion.

Note that if an end distance of twice the diameter of a bolt is used when detailing it always exceeds both values shown in Table 7, and therefore takes care of end and edge distance simultaneously.

Standard spacing of holes in structural sections
Details for standard spacing of holes in universal columns, joists, universal beams, channels and angles are given in *The Structural Steelwork Handbook*, published by BCSA/CONSTRADO. An expanded form of this information is shown in Tables 8–11 and Fig. 30. These dimensions comply with the rules previously specified for spacing and edge distance for holes.

Leg length (mm)	Back marks and max. dia. of bolt
200	75, 55, 75, 30 mm dia. / 55, 55, 55, 20 mm dia.
150	55, 55, 20 mm dia.
125	45, 50, 20 mm dia.
120	45, 50, 16 mm dia.
100	55, 24 mm dia.

Leg length (mm)	Back marks and max. dia. of bolt
90	50, 24 mm dia.
80 / 75	45, 20 mm dia.
70	40, 20 mm dia.
65	35, 20 mm dia.
60	35, 16 mm dia.
50	28, 12 mm dia.

Fig. 30. *Spacing of holes in angles.*

TABLE 8. SPACING OF HOLES IN FLANGES OF UNIVERSAL COLUMNS

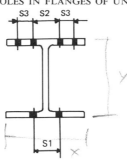

Serial size	Mass	Actual flange width B	Standard spacing (mm)			Recommended dia of bolt (mm)	
x y			S1	S2	S3	2 holes	4 holes
356 × 406	634	424.1	140	140	75	24	24
	551	418.5					
	467	412.4					
	393	407.0					
	340	403.0					
	287	399.0					
	235	395.0					
Column core	477	424.4	140	140	75	24	24
356 × 368	202	374.4	140	140	75	24	24
	177	372.1					
	153	370.2					
	129	368.3					
305 × 305	283	321.8	140	120	60	24	24
	240	317.9					
	198	314.1					
	158	310.6	140	120	60	24	20
	137	308.7					
	118	306.8					
	97	304.8					
254 × 254	167	264.5	140			24	
	132	261.0					
	107	258.3					
	89	255.9					
	73	254.0					
203 × 203	86	208.8	140			20	
	71	206.2					
	60	205.2					
	52	203.9					
	46	203.2					
152 × 152	37	154.4	90			20	
	30	152.9					
	23	152.4					

TABLE 9. SPACING OF HOLES IN FLANGES OF JOISTS

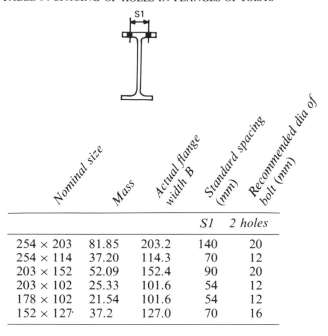

Nominal size	Mass	Actual flange width B	Standard spacing (mm)	Recommended dia of bolt (mm)
			S1	2 holes
254 × 203	81.85	203.2	140	20
254 × 114	37.20	114.3	70	12
203 × 152	52.09	152.4	90	20
203 × 102	25.33	101.6	54	12
178 × 102	21.54	101.6	54	12
152 × 127·	37.2	127.0	70	16

TABLE 10. SPACING OF HOLES IN FLANGES OF UNIVERSAL BEAMS

Serial size	Mass	Actual flange width B	Standard spacing (mm)			Recommended dia of bolt (mm)	
			S1	S2	S3	2 holes	4 holes
914 × 419	388	420.5	140	140	75	24	24
	343	418.5					
914 × 305	289	307.8	140	120	60	24	20
	253	305.5					
	224	304.1					
	201	303.4					
838 × 292	226	293.8	140			24	
	194	292.4					
	176	291.6					

Serial size	Mass	Actual flange width B	Standard spacing (mm)			Recommended dia of bolt (mm)	
			S1	S2	S3	2 holes	4 holes
762 × 267	197	268.0	140			24	
	173	266.7					
	147	265.3					
686 × 254	170	255.8	140			24	
	152	254.5					
	140	253.7					
	125	253.0					
610 × 305	238	311.6	140	120	60	24	20
	179	307.0					
	149	304.8					
610 × 229	140	230.1	140			24	
	125	229.0					
	113	228.2					
	101	227.6					
533 × 210	122	211.9	140			20	
	109	210.7					
	101	210.1					
	92	209.3					
	82	208.7					
457 × 191	98	192.8	90			24	
	89	192.0					
	82	191.3					
	74	190.5					
	67	189.9					
457 × 152	82	153.5	90			20	
	74	152.7					
	67	151.9					
	60	152.9					
	52	152.4					
406 × 178	74	179.7	90			24	
	67	178.8					
	60	177.8					
	54	177.6					
406 × 140	46	142.4	70			20	
	39	141.8					
356 × 171	67	173.2	90			24	
	57	172.1					
	51	171.5					
	45	171.0					
356 × 127	39	126.0	70			16	
	33	125.4					

Serial size	Mass	Actual flange width B	Standard spacing (mm)			Recommended dia of bolt (mm)	
			S1	S2	S3	2 holes	4 holes
305 × 165	54	166.8	90			24	
	46	165.7					
	40	165.1					
305 × 127	48	125.2	70			16	
	42	124.3					
	37	123.5					
305 × 102	33	102.4	54			12	
	28	101.9					
	25	101.6					
254 × 146	43	147.3	70			20	
	37	146.4					
	31	146.1					
254 × 102	28	102.1	54			12	
	25	101.9					
	22	101.6					
203 × 133	30	133.8	70			20	
	25	133.4					

TABLE 11. SPACING OF HOLES IN FLANGES OF
CHANNELS*

Back mark (B.M.)

Nominal flange width (mm)	(mm)	Recommended dia of bolt (mm)
102	55	24
89	55	20
76	45	20
64	35	16
51	30	10

*Back mark dimension depends on the nominal flange width.

Machining

The detailer should understand the significance of machining (or not machining) bearing surfaces at bases, caps, splices and butt joints.

Machining is carried out over the bearing surfaces at the ends and joints of compression members dependent on direct contact for the transmission of compressive stresses from steel to steel. Instructions regarding any machining should be noted on the relevant detail drawings. Four important types of connections are considered here and are illustrated in Fig. 31:

(a) stanchion bases, gusseted and slab types;
(b) stanchion caps;
(c) stanchion splices;
(d) butt joints in compression members.

Gusseted base
For stanchions with gusseted bases the gusset plates, angle cleats, stiffeners, fastenings, etc., in combination with the end bearing area of the stanchion shaft should, when all parts are machined flush for bearing, be adequate to transmit the loads, reactions and bending moments to the base plate without exceeding allowable stresses.

When the end of the stanchion shaft and the gusset plates are not machined flush for complete bearing the fastenings connecting them to the base plate should adequately transmit all the forces to which the base is subjected.

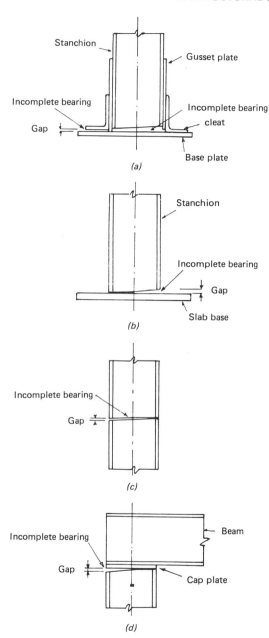

Fig. 31. *Four types of connection.* (*a*) Gusseted base. (*b*) Slab base. (*c*) Stanchion splice. Splice plates not shown. (*d*) Stanchion cap plate.

Slab bases
Fastenings should be provided to retain the parts securely in place and to resist adequately all forces and moments other than direct com- pression. The stanchion end and the top of the slab base should be machined for complete bearing.

Stanchion caps
The underside of the cap plate and the end of the stanchion shaft should be machined for com- plete bearing; otherwise fasteners and connec- tions should adequately transfer the applied loads.

Stanchion splices
At joints where both ends of compression members are machined for bearing over the whole area a splice capable of holding the members correctly in place and resisting any tension due to bending is sufficient. If the ends are not machined the joint should be designed to transmit all applied loads and moments.

Butt joints
At a butt joint in a compression member, e.g. a top boom of a lattice girder, both ends should be machined for complete bearing. If the ends are not machined then the joint should be designed to transmit all applied loads and moments.

Painting
Instructions regarding the painting of the steel- work should be given with the general notes on each detail drawing. All surfaces to be painted should be dry and all loose scale and rust removed.

Contact surfaces
It is not necessary for shop contact surfaces to be painted unless so specified. However if so specified they should be brought together while the paint is still wet.

Inaccessible surfaces
Surfaces not in contact, but inaccessible after shop assembly of the steelwork, should receive the specified treatment before assembling takes place.

Encased steel
Steelwork to be encased in concrete should not be painted.

Site painting
Painting instructions required after erection of the steelwork on the site should be noted separately.

TABLE 12. DIMENSIONS FOR DETAILING UNIVERSAL BEAMS (PARALLEL FLANGES)*

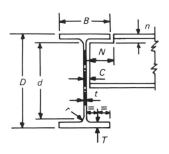

| Serial size (mm) | Mass per metre (kg) | Depth of section D (mm) | Root radius r (mm) | Flange | | Web | | End clearance | Notch | |
				Width B (mm)	Thickness T (mm)	Thickness t (mm)	Depth d (mm)	C (mm)	N (mm)	n (mm)
914 × 419	388	920.5	24.1	420.5	36.6	21.5	799.0	13	210	62
	343	911.4	24.1	418.5	32.0	19.4	799.0	12	210	58
914 × 305	289	926.6	19.1	307.8	32.0	19.6	824.4	12	156	52
	253	918.5	19.1	305.5	27.9	17.3	824.4	11	156	48
	224	910.3	19.1	304.1	23.9	15.9	824.4	10	156	44
	201	903.0	19.1	303.4	20.2	15.2	824.4	10	156	40
838 × 292	226	850.9	17.8	293.8	26.8	16.1	761.7	10	150	46
	194	840.7	17.8	292.4	21.7	14.7	761.7	9	150	40
	176	834.9	17.8	291.6	18.8	14.0	761.7	9	150	38
762 × 267	197	769.6	16.5	268.0	25.4	15.6	685.8	10	138	42
	173	762.0	16.5	266.7	21.6	14.3	685.8	9	138	40
	147	753.9	16.5	265.3	17.5	12.9	685.8	8	138	36
686 × 254	170	692.9	15.2	255.8	23.7	14.5	615.0	9	132	40
	152	687.6	15.2	254.5	21.0	13.2	615.0	9	132	38
	140	683.5	15.2	253.7	19.0	12.4	615.0	8	132	36
	125	677.9	15.2	253.0	16.2	11.7	615.0	8	132	32
610 × 305	238	633.0	16.5	311.5	31.4	18.6	537.2	11	158	48
	179	617.5	16.5	307.0	23.6	14.1	537.2	9	158	42
	149	609.6	16.5	304.8	19.7	11.9	537.2	8	158	38
610 × 229	140	617.0	12.7	230.1	22.1	13.1	547.1	9	120	36
	125	611.9	12.7	229.0	19.6	11.9	547.1	8	120	34
	113	607.3	12.7	228.2	17.3	11.2	547.1	8	120	32
	101	602.2	12.7	227.6	14.8	10.6	547.1	7	120	28

Serial size	Mass per metre	Depth of section D	Root radius r	Flange		Web		End clearance	Notch	
				Width B	Thickness T	Thickness t	Depth d	C	N	n
(mm)	(kg)	(mm)	(mm)	(mm)	(mm)	(mm)	(mm)	(mm)	(mm)	(mm)
533 × 210	122	544.6	12.7	211.9	21.3	12.8	476.5	8	110	34
	109	539.5	12.7	210.7	18.8	11.6	476.5	8	110	32
	101	536.7	12.7	210.1	17.4	10.9	476.5	7	110	30
	92	533.1	12.7	209.3	15.6	10.2	476.5	7	110	30
	82	528.3	12.7	208.7	13.2	9.6	476.5	7	110	26
457 × 191	98	467.4	10.2	192.8	19.6	11.4	407.9	8	102	30
	89	463.6	10.2	192.0	17.7	10.6	407.9	7	102	28
	82	460.2	10.2	191.3	16.0	9.9	407.9	7	102	28
	74	457.2	10.2	190.5	14.5	9.1	407.9	7	102	26
	67	453.6	10.2	189.9	12.7	8.5	407.9	6	102	24
457 × 152	82	465.1	10.2	153.5	18.9	10.7	406.9	7	82	30
	74	461.3	10.2	152.7	17.0	9.9	406.9	7	82	28
	67	457.2	10.2	151.9	15.0	9.1	406.9	7	82	26
	60	454.7	10.2	152.9	13.3	8.0	407.7	6	82	24
	52	449.8	10.2	152.4	10.9	7.6	407.7	6	82	22
406 × 178	74	412.8	10.2	179.7	16.0	9.7	360.5	7	96	28
	67	409.4	10.2	178.8	14.3	8.8	360.5	6	96	26
	60	406.4	10.2	177.8	12.8	7.8	360.5	6	96	24
	54	402.6	10.2	177.6	10.9	7.6	360.5	6	96	22
406 × 140	46	402.3	10.2	142.4	11.2	6.9	359.6	5	78	22
	39	397.3	10.2	141.8	8.6	6.3	359.6	5	78	20
356 × 171	67	364.0	10.2	173.2	15.7	9.1	312.2	7	94	26
	57	358.6	10.2	172.1	13.0	8.0	312.2	6	94	24
	51	355.6	10.2	171.5	11.5	7.3	312.2	6	94	22
	45	352.0	10.2	171.0	9.7	6.9	312.2	5	94	20
356 × 127	39	352.8	10.2	126.0	10.7	6.5	311.1	5	70	22
	33	348.5	10.2	125.4	8.5	5.9	311.1	5	70	20
305 × 165	54	310.9	8.9	166.8	13.7	7.7	265.6	6	90	24
	46	307.1	8.9	165.7	11.8	6.7	265.6	5	90	22
	40	303.8	8.9	165.1	10.2	6.1	265.6	5	90	20
305 × 127	48	310.4	8.9	125.2	14.0	8.9	264.6	6	70	24
	42	306.6	8.9	124.3	12.1	8.0	264.6	6	70	22
	37	303.8	8.9	123.5	10.7	7.2	264.6	6	70	20

Serial size (mm)	Mass per metre (kg)	Depth of section D (mm)	Root radius r (mm)	Flange Width B (mm)	Flange Thickness T (mm)	Web Thickness t (mm)	Web Depth d (mm)	End clearance C (mm)	Notch N (mm)	Notch n (mm)
305 × 102	33	312.7	7.6	102.4	10.8	6.6	275.8	5	58	20
	28	308.9	7.6	101.9	8.9	6.1	275.8	5	58	18
	25	304.8	7.6	101.6	6.8	5.8	275.8	5	58	16
254 × 146	43	259.6	7.6	147.3	12.7	7.3	218.9	6	80	22
	37	256.0	7.6	146.4	10.9	6.4	218.9	5	80	20
	31	251.5	7.6	146.1	8.6	6.1	218.9	5	80	18
254 × 102	28	260.4	7.6	102.1	10.0	6.4	225.0	5	58	18
	25	257.0	7.6	101.9	8.4	6.1	225.0	5	58	16
	22	254.0	7.6	101.6	6.8	5.8	225.0	5	58	16
203 × 133	30	206.8	7.6	133.8	9.6	6.3	172.3	5	74	18
	25	203.2	7.6	133.4	7.8	5.8	172.3	5	74	16

The dimension N is based upon the outstand from web face to flange edge + 10 mm to nearest 2 mm above, and makes due allowance for rolling tolerance.

The dimension $n = (D - d)/2$ to the nearest 2 mm above.

The dimension $C = t/2 + 2$ mm to nearest mm.

* From *Metric Practice for Structural Steelwork* (BCSA).

TABLE 13. DIMENSIONS FOR DETAILING UNIVERSAL COLUMNS*

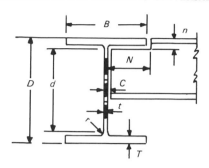

Serial size	Mass per metre	Depth of section D	Root radius r	Flange Width B	Flange Thickness T	Web Thickness t	Web Depth d	End clearance C	Notch N	Notch n
(mm)	(kg)	(mm)	(mm)	(mm)	(mm)	(mm)	(mm)	(mm)	(mm)	(mm)
356 × 406	634	474.7	15.2	424.1	77.0	47.6	290.1	26	200	94
	551	455.7	15.2	418.5	67.5	42.0	290.1	23	200	84
	467	436.6	15.2	412.4	58.0	35.9	290.1	20	200	74
	393	419.1	15.2	407.0	49.2	30.6	290.1	17	200	66
	340	406.4	15.2	403.0	42.9	26.5	290.1	15	200	60
	287	393.7	15.2	399.0	36.5	22.6	290.1	13	200	52
	235	381.0	15.2	395.0	30.2	18.5	290.1	11	200	46
Column Core	477	427.0	15.2	424.4	53.2	48.0	290.1	26	200	70
356 × 368	202	374.7	15.2	374.4	27.0	16.8	290.1	10	190	44
	177	368.3	15.2	372.1	23.8	14.5	290.1	9	190	40
	153	362.0	15.2	370.2	20.7	12.6	290.1	8	190	36
	129	355.6	15.2	368.3	17.5	10.7	290.1	7	190	34
305 × 305	283	365.3	15.2	321.8	44.1	26.9	246.5	15	158	60
	240	352.6	15.2	317.9	37.7	23.0	246.5	13	158	54
	198	339.9	15.2	314.1	31.4	19.2	246.5	12	158	48
	158	327.2	15.2	310.6	25.0	15.7	246.5	10	158	42
	137	320.5	15.2	308.7	21.7	13.8	246.5	9	158	38
	118	314.5	15.2	306.8	18.7	11.9	246.5	8	158	34
	97	307.8	15.2	304.8	15.4	9.9	246.5	7	158	32
254 × 254	167	289.1	12.7	264.5	31.7	19.2	200.2	12	134	46
	132	276.4	12.7	261.0	25.1	15.6	200.2	10	134	30
	107	266.7	12.7	258.3	20.5	13.0	200.2	9	134	34
	89	260.4	12.7	255.9	17.3	10.5	200.2	7	134	32
	73	254.0	12.7	254.0	14.2	8.6	200.2	6	134	28

Serial size	Mass per metre	Depth of section D	Root radius r	Flange		Web		End clearance	Notch	
				Width B	Thickness T	Thickness t	Depth d	C	N	n
(mm)	(kg)	(mm)	(mm)	(mm)	(mm)	(mm)	(mm)	(mm)	(mm)	(mm)
203 × 203	86	222.3	10.2	208.8	20.5	13.0	160.8	8	108	32
	71	215.9	10.2	206.2	17.3	10.3	160.8	7	108	28
	60	209.6	10.2	205.2	14.2	9.3	160.8	7	108	26
	52	206.2	10.2	203.9	12.5	8.0	160.8	6	108	24
	46	203.2	10.2	203.2	11.0	7.3	160.8	6	108	22
152 × 152	37	161.8	7.6	154.4	11.5	8.1	123.4	6	84	20
	30	157.5	7.6	152.9	9.4	6.6	123.4	5	84	18
	23	152.4	7.6	152.4	6.8	6.1	123.4	5	84	16

The dimension N is based upon the outstand from web face to flange edge + 10 mm to nearest 2 mm above, and makes due allowance for rolling tolerance.

The dimension $n = (D - d)/2$ to the nearest 2 mm above.

The dimension $C = t/2 + 2$ mm to the nearest mm.

* From *Metric Practice for Structural Steelwork* (BCSA).

TABLE 14. DIMENSIONS FOR DETAILING JOISTS*

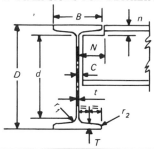

Nominal size	Mass per metre	Depth of section	Root radius	Toe radius	Flange		Inside slope	Web Thickness	End clearance Depth		Notch	
		D	r_1	r_2	Width B	Thickness T		t	d	C	N	n
(mm)	(kg)	(mm)	(mm)	(mm)	(mm)	(mm)	(degrees)	(mm)	(mm)	(mm)	(mm)	(mm)
254 × 203	*81.85*	254.0	19.6	9.7	203.2	19.9	8	10.2	166.6	7	102	44
254 × 114	*37.20*	254.0	12.4	6.1	114.3	12.8	8	7.6	199.2	6	60	28
203 × 152	*52.09*	203.2	15.5	7.6	152.4	16.5	8	8.9	133.2	7	78	36
203 × 102	25.33	203.2	9.4	3.2	101.6	10.4	5	5.8	161.0	5	54	22
178 × 102	21.54	177.8	9.4	3.2	101.6	9.0	5	5.3	138.2	5	54	20
152 × 127	*37.20*	152.4	13.5	6.6	127.0	13.2	8	10.4	94.3	7	64	30
152 × 89	17.09	152.4	7.9	2.4	88.9	8.3	5	4.9	117.7	5	48	18
152 × 76	*17.86*	152.4	9.4	4.6	76.2	9.6	8	5.8	111.9	5	40	22
127 × 114	*29.76*	127.0	9.9	4.8	114.3	11.5	8	10.2	79.4	7	58	24
127 × 114	26.79	127.0	9.9	5.0	114.3	11.4	8	7.4	79.5	6	60	24
127 × 76	*16.37*	127.0	9.4	4.6	76.2	9.6	8	5.6	86.5	5	42	22
127 × 76	13.36	127.0	7.9	2.4	76.2	7.6	5	4.5	94.2	4	42	18
114 × 114	*26.79*	114.3	14.3	2.4	114.3	10.7	8	9.5	60.8	7	58	28
102 × 102	*23.07*	101.6	11.1	3.2	101.6	10.3	8	9.5	55.1	7	52	24
102 × 64	9.65	101.6	6.9	2.4	63.5	6.6	5	4.1	73.2	4	36	16
102 × 44	*7.44*	101.6	6.9	3.3	44.4	6.1	8	4.3	74.7	4	26	14
89 × 89	*19.35*	88.9	11.1	3.2	88.9	9.9	8	9.5	44.1	7	46	24
76 × 76	*14.67*	76.2	9.5	3.2	80.0	8.4	8	8.9	38.0	7	42	20
76 × 76	*12.65*	76.2	9.4	4.6	76.2	8.4	8	5.1	37.9	5	42	20

Sections with mass shown in italics are, although frequently rolled, not in BS4. Availability should be checked with BSC Sections Product Unit.

The dimension N is equal to the outstand from web face to flange edge + 6 mm to nearest 2 mm above.

The dimension $n = (D - d)/2$ to nearest 2 mm above.

The dimension $C = t/2 + 2$ mm to the nearest mm.

* From *Metric Practice for Structural Steelwork* (BCSA).

TABLE 15. DIMENSIONS FOR DETAILING CHANNELS*

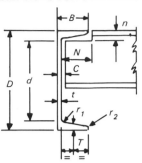

Nominal size	Mass per metre	Depth of section	Root radius	Toe radius	Flange		Inside slope	Web	End clearance		Notch	
					Width	Thick-ness		Thick-ness	Depth			
		D	r_1	r_2	B	T		t	d	C	N	n
(mm)	(kg)	(mm)	(mm)	(mm)	(mm)	(mm)	(degrees)	(mm)	(mm)	(mm)	(mm)	(mm)
432 × 102	65.54	431.8	15.2	4.8	101.6	16.8	5	12.2	362.5	14	96	36
381 × 102	55.10	381.0	15.2	4.8	101.6	16.3	5	10.4	312.4	12	98	36
305 × 102	46.18	304.8	15.2	4.8	101.6	14.8	5	10.2	239.3	12	98	34
304 × 89	41.69	304.8	13.7	3.2	88.9	13.7	5	10.2	245.4	12	86	30
254 × 89	35.74	254.0	13.7	3.2	88.9	13.6	5	9.1	194.8	11	86	30
254 × 76	28.29	254.0	12.2	3.2	76.2	10.9	5	8.1	203.7	10	76	26
229 × 89	32.76	228.6	13.7	3.2	88.9	13.3	5	8.6	169.9	11	88	30
229 × 76	26.06	228.6	12.2	3.2	76.2	11.2	5	7.6	178.1	10	76	26
203 × 89	29.78	203.2	13.7	3.2	88.9	12.9	5	8.1	145.3	10	88	30
203 × 76	23.82	203.2	12.2	3.2	76.2	11.2	5	7.1	152.4	9	76	26
178 × 89	26.81	177.8	13.7	3.2	88.9	12.3	5	7.6	120.9	10	88	30
178 × 76	20.84	177.8	12.2	3.2	76.2	10.3	5	6.6	128.8	9	76	26
152 × 89	23.84	152.4	13.7	3.2	88.9	11.6	5	7.1	97.0	9	88	28
152 × 76	17.88	152.4	12.2	2.4	76.2	9.0	5	6.4	105.9	8	76	24
127 × 44	14.90	127.0	10.7	2.4	63.5	9.2	5	6.4	84.1	8	64	22
102 × 51	10.42	101.6	9.1	2.4	50.8	7.6	5	6.1	65.8	8	52	18
76 × 38	6.70	76.2	7.6	2.4	38.1	6.8	5	5.1	45.7	7	40	16

The dimension N is equal to the outstand from web face to flange edge + 6 mm to nearest 2 mm above.
The dimension $n = (D - d)/2$ to nearest 2 mm above.
The dimension $C = t + 2$ mm to the nearest mm.
* From *Metric Practice for Structural Steelwork* (BCSA).

TABLE 16. DIMENSION OF ANGLES FOR DETAILING*

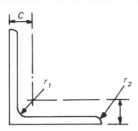

Equal angles					Unequal angles					
Standard size	Standard thickness	Root radius	Toe radius	Centre of gravity	Standard size	Standard thickness	Root radius	Toe radius	Centre of gravity	
		r_1	r_2	C	$A \times B$		r_1	r_2	Cx	Cy
(mm)	(mm)	(mm)	(mm)	(mm)	(mm)	(mm)	(mm)	(mm)	(mm)	(mm)
25 × 25	3	3.5	2.4	7.2	65 × 50	5	6.0	2.4	19.9	12.5
	4			7.6		6			20.4	12.9
	5			8.0		8			21.1	13.7
40 × 40	4	6.0	2.4	11.2						
	5			11.6	75 × 50	6	7.0	2.4	24.4	12.1
	6			12.0		8			25.2	12.9
45 × 45	4	7.0	2.4	12.3						
	5			12.8						
	6			13.2	80 × 60	6	8.0	4.8	24.7	14.8
						7			25.1	15.2
50 × 50	5	7.0	2.4	14.0		8			25.5	15.6
	6			14.5						
	8			15.2						
					100 × 65	7	10.0	4.8	32.3	15.1
60 × 60	5	8.0	2.4	16.4		8			32.7	15.5
	6			16.9		10			33.6	16.3
	8			17.7						
	10			18.5						
70 × 70	6	9.0	2.4	19.3	100 × 75	8	10.0	4.8	31.0	18.7
	8			20.1		10			31.9	19.5
	10			20.9		12			32.7	20.3
80 × 80	6	10.0	4.8	21.7						
	8			22.6	125 × 75	8	11.0	4.8	41.4	16.8
	10			23.4		10			42.3	17.6
						12			43.1	18.4
90 × 90	6	11.0	4.8	24.1						
	8			25.0						
	10			25.8						
	12			26.6	150 × 75	10	11.0	4.8	53.2	16.1

Equal angles					Unequal angles					
Standard size	Standard thickness	Root radius	Toe radius	Centre of gravity	Standard size	Standard thickness	Root radius	Toe radius	Centre of gravity	
		r_1	r_2	C	$A \times B$		r_1	r_2	Cx	Cy
(mm)	(mm)	(mm)	(mm)	(mm)	(mm)	(mm)	(mm)	(mm)	(mm)	(mm)
						12			54.1	16.9
100 × 100	8	12.0	4.8	27.4		15			55.3	18.1
	12			29.0						
	15			30.2						
					150 × 90	10	12.0	4.8	50.0	20.4
120 × 120	8	13.0	4.8	32.3		12			50.8	21.2
	10			33.1		15			52.1	22.3
	12			34.0						
	15			35.1						
150 × 150	10	16.0	4.8	40.3	200 × 100	10	15.0	4.8	69.3	20.1
	12			41.2		12			70.3	21.0
	15			42.5		15			71.6	22.2
	18			43.7						
200 × 200	16	18.0	4.8	55.2						
	18			56.0	200 × 150	12	15.0	4.8	60.8	36.1
	20			56.8		15			62.1	37.3
	24			58.4		18			63.3	38.5

*From *Metric Practice for Structural Steelwork* (BCSA).

SELF-ASSESSMENT QUESTIONS

1. Explain what is meant by the following terms: (a) holing; (b) edge distance; (c) end distance; (d) pitch; (e) gauge lines; (f) gauge; (g) cross centres; (h) back mark.

2. How are the following structural steel sections designated: (a) universal columns; (b) joists; (c) channels, (d) angles; (e) rectangular hollow sections.

3. Explain the importance of erection or identification marks given to structural steel members.

4. Fig. 32 shows three typical connections, (a), (b), and (c), where notching of beams is required. Determine the notch dimensions N, n, n_1 and n_2 where appropriate.

5. Fig. 33 shows two typical connections, (a) and (b). Fill in the missing dimensions and show the maximum diameter of holes that may be used.

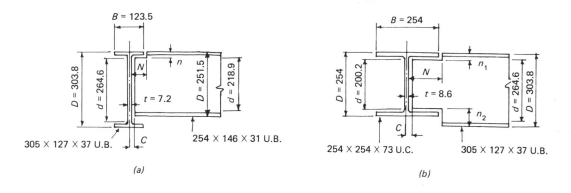

(a)

(b)

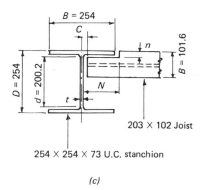

(c)

Fig. 32. (a) *Elevation.* (b) *Elevation.* (c) *Plan.*

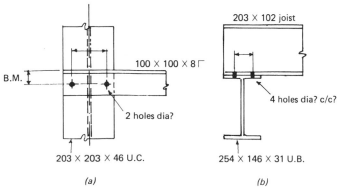

(a)

(b)

Fig. 33. (a) *Elevation.* (b) *Elevation.*

CHAPTER FOUR

Framing Plans and Stanchion Schedules

CHAPTER OBJECTIVES
After studying this chapter you should be able to:
• prepare general layout framing plans;
• prepare stanchion schedules;
• prepare material requisition lists for steel framed structures.

INTRODUCTION

In Chapter Three reference was made to the standard practice of placing identification or erection marks on all members of the steel frame for detailing, fabrication and erection purposes.

Various systems of notation are used by structural steel fabricators and some typical systems are described and illustrated in this chapter. Assembly marks which are introduced and applied to the shop detail drawings are dealt with in subsequent chapters.

Requisition lists for ordering and allocating material for the main structural members are prepared from the general layout framing plans.

Identification or erection marks will now be applied to a variety of structures, both of single storey and multi-storey construction, using various systems of notation.

SINGLE STOREY BUILDINGS

Three common types of construction are considered:

(a) flat roof;
(b) pitched roof with trusses;
(c) pitched roof with portal frames.

Each of these types will be now described and illustrated.

Flat roof
An example is illustrated in Fig. 34. This is a flat roof deck carried by horizontal steel beams supported at their ends by vertical steel stanchions. Four schemes are considered.

Scheme 1
Stanchions are indicated by I or H, depending on their orientation on the design drawings and are numbered (1), etc., starting at the left hand bottom corner of the drawing and working left to right for each row of stanchions. They may be finally prefixed S, e.g. (S1), (S2), etc. Beam marks generally bear some relationship to the numbers or marks of the stanchions.

The beam mark consists of the stanchion number or mark as a prefix, together with the addition of even numbers 2, 4, etc., for beams shown running left to right on the plan and odd numbers 1, 3, etc., for beams shown running from the bottom to the top on the plan.

For example, starting at stanchion (1) and reading from left to right, the first beam mark 12 receives its first number 1 from the stanchion and its second number is an even number 2. From stanchion (2) and reading left to right, the beam mark 22 receives its first number 2 from the stanchion and its second number is an even number 2. Beam mark 24 has no stanchion at the left hand end but is related to stanchion (2) and receives its first number 2 from the stanchion, its second number being the even number 4.

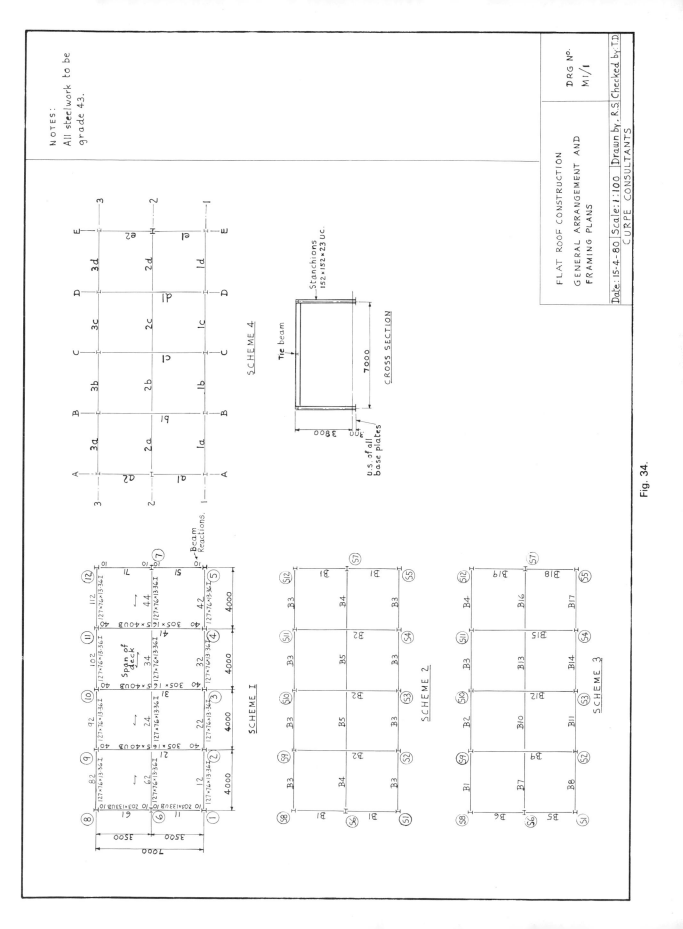

Fig. 34.

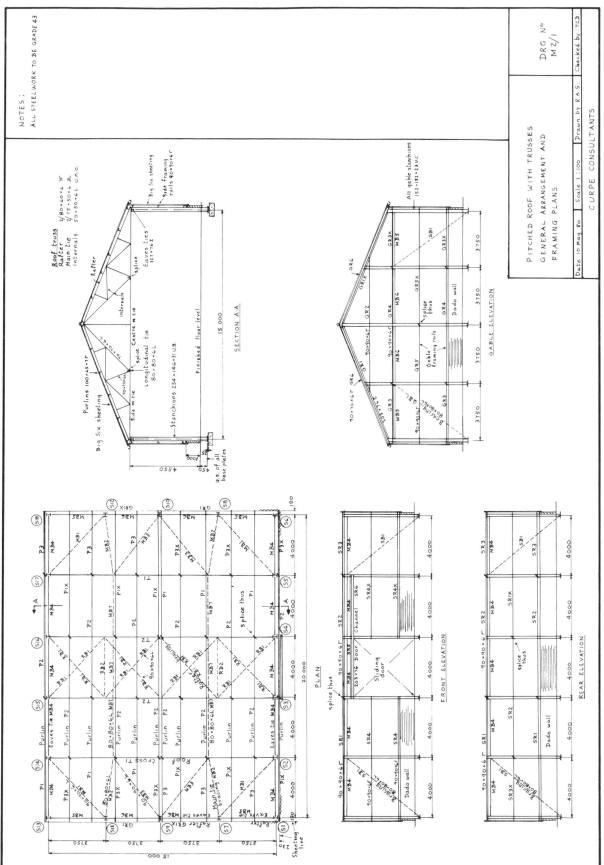

Fig. 35.

Again starting from stanchion ①︎ and reading from bottom to top, the beam mark 11 receives the number 1 from the stanchion and its second number is an odd number 1. From stanchion ②︎, and reading from bottom to top, the beam mark 21 receives the number 2 from the stanchion and its second number is the odd number 1.

Beams may be prefixed B, e.g., B12, B11, etc.

Scheme 2
Stanchions are shown and given marks Ⓢ①, Ⓢ②, etc. Beams are marked B1, B2, etc. All beams which are identical in every respect (including connections) are given the same mark.

Scheme 3
Stanchions are shown and given marks Ⓢ①, Ⓢ②, etc. Each beam, whether identical to any other or not, is given a different mark, e.g. B1, B2, B3, etc.

Scheme 4
Stanchions and beams are given marks relative to grid lines. Grid lines are shown as letters A, B, C, etc., in one direction and as numbers 1, 2, 3, etc., in the other direction.

Stanchion marks are related to the intersection of the grid lines, e.g. A3 where grid lines A and 3 intersect.

Beam marks are related to the grid lines. Reading from left to right along grid line 1, beams are marked 1a, 1b, 1c, 1d, etc. Reading from bottom to top along grid line A, beams are marked a1, a2, a3, etc.

Pitched roof with trusses

An example is illustrated in Fig. 35.

Stanchions are marked Ⓢ①, Ⓢ②, etc. Generally the remainder of the members which are identical in every respect are given the same mark. Roof trusses are marked T1, T2, etc.

Both trusses marked T1 are identical in every respect and are therefore given the same mark.

Both trusses marked T2 have additional connections for rafter bracing and are therefore given a different mark to T1.

Main tie bracing is marked MB1, MB2, etc.

Rafter bracing is marked RB1, RB2, etc.

Purlins are marked P1, P2, etc., and the splices or joints should be staggered as shown. Purlins bearing the mark P1X are opposite hand to those marked P1, i.e. a left and a right (*see* Fig. 36).

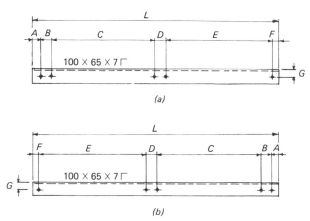

Fig. 36. *Opposite handed purlins.* (*a*) Purlin P1. (*b*) Purlin P1X.

Side framing rails at the front and rear elevations are marked SR1, SR2, etc.

Gable framing rails at the gables are marked GR1, GR2, etc.

Pitched roof with portal frames

An example is shown in Fig. 37, illustrating alternative systems.

Stanchions or portal legs are marked Ⓢ①, Ⓢ②, etc., or alternatively, using grid lines, A1, B1, etc. The remainder of the structure is similar for both systems.

Portal rafters are marked PR1, PR2, etc. Portal rafters PR1 are identical in every respect (including connections) and are therefore given the same mark. Rafters PR2 have additional connections provided for rafter bracing and are therefore given a different mark to PR1.

Eaves ties are marked B1.

Rafter bracing is marked RB1.

Purlins are marked P1, P2, etc., as for a pitched roof with trusses.

MULTI-STOREY BUILDINGS

An example is illustrated in Fig. 38.

Stanchions are numbered ①︎, ②︎, etc.,

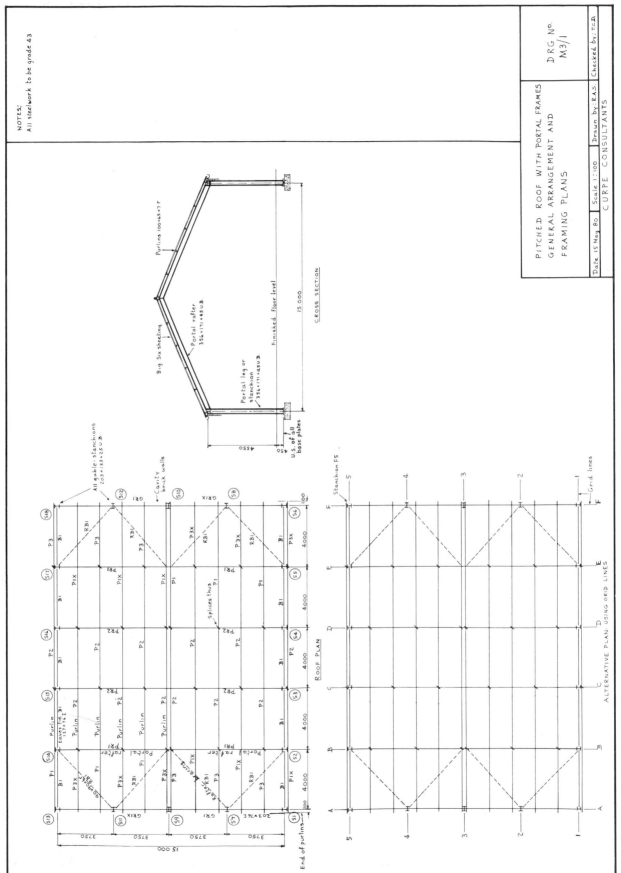

Fig. 37

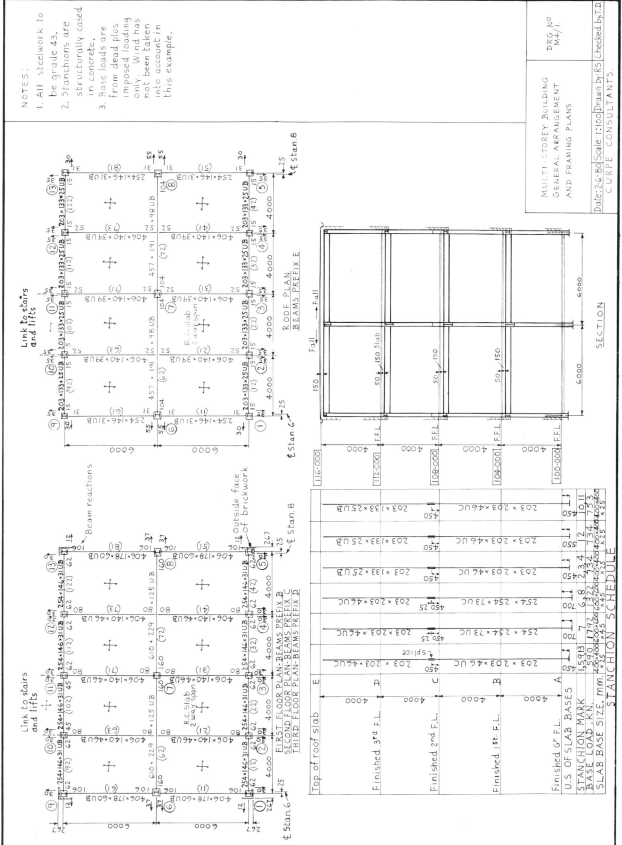

Fig. 38.

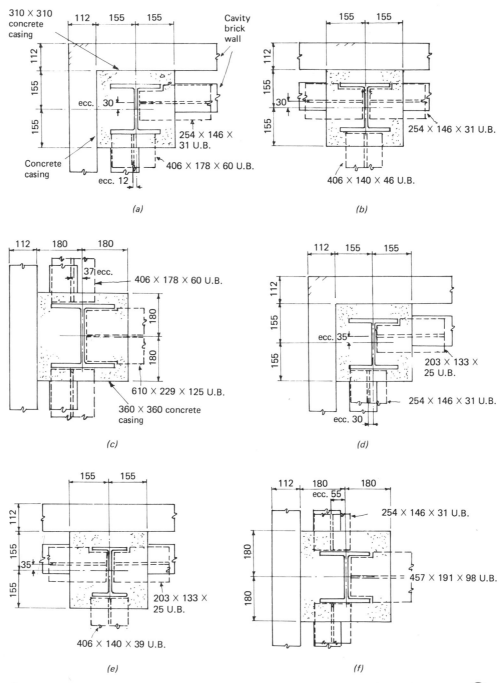

Fig. 39. *Beam eccentricities determined at the external stanchions for Fig. 38.* (*a*) Stanchions ①, ⑤, ⑨, ⑬—floor level. (*b*) Stanchions ②, ③, ④, ⑩, ⑪, ⑫—floor level. (*c*) Stanchions ⑥, ⑧—floor level. (*d*) Stanchions ①, ⑤, ⑨, ⑬—roof level. (*e*) Stanchions ②, ③, ④, ⑩, ⑪, ⑫—roof level. (*f*) Stanchions ⑥, ⑧—roof level. (Scale 1:10.)

starting from the bottom left hand corner of the plan, and are finally prefixed S, e.g. (S1), (S2), etc., to complete the marking. This system for the stanchions is identical to scheme 1 for a flat-roofed single storey building.

Floor levels are lettered A, B, C, D for ground, first, second and third respectively, and E for the roof level.

Beam marks are shown on plans as (12), (22), (11), (61), etc. These marks are determined in the same manner as those given in scheme 1 for flat-roofed single storey buildings. To complete the marking, beams at first-floor level, i.e. level B, are prefixed B, e.g. (B12), (B22), (B11), (B61), etc.

All floor beams and roof beams are treated in a similar manner and prefixed by the appropriate letter.

Section size of beams and end reactions are indicated on the plan to enable the detailer to design and provide adequate beam to stanchion and beam to beam connections.

DETERMINATION OF ECCENTRICITIES

External wall beams for this building, are generally eccentric to the stanchions, i.e. the centre lines of the beams do not coincide with the centre lines of the stanchions. These eccentricities are determined from the relative positions and dimensions of the brick walls, the concrete casing and the stanchion and beam section sizes, and are shown as 35↑ (the arrow indicates the direction and the number the magnitude of the eccentricity). Suggested minimum dimensions for concrete casing to structurally cased stanchions are taken from the BCSA/CONSTRADO *Structural Steelwork Handbook* and are shown as 310 × 310 and 360 × 360 on Fig. 39.

In this example cover to the casing by the outer leaf of brickwork is taken as 112 mm including mortar joint thickness. The outside edges of the beam flanges are positioned approximately flush with the outside edges of the external stanchions. Figure 39 illustrates the beam eccentricities determined at the external stanchions, taking into account the above factors, while Fig. 40 illustrates the concrete casing to the external beams and the lightweight casing to the internal beams.

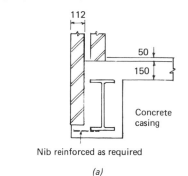

Nib reinforced as required

(a)

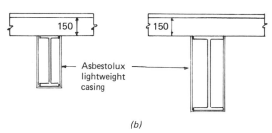

(b)

Fig. 40. *(a) Concrete casing to external beams. (b) Lightweight casing to internal beams.*

STANCHION SCHEDULE

When all the stanchions in a structure have been designed it is desirable to present them in the form of a single drawing. A stanchion schedule gathers together a good deal of relevant information into a suitable condensed form.

A stanchion schedule for the multi-storey building example is shown in Fig. 38, giving:

(a) stanchion marks;
(b) vertical dimensions between floors;
(c) depth to underside (us) of slab bases;
(d) stanchion base loads;
(e) slab base dimensions;
(f) splice positions;
(g) stanchion section sizes.

Stanchions possessing identical section sizes, base loads, slab base dimensions and base levels are grouped together. Note that the only difference between stanchion 12 and stanchions 2, 3 and 4 is the depth to the underside of the slab base and it is therefore shown separately.

Note also that the dimension 25 shown at the splice for stanchions 6, 7 and 8 would not normally appear but is given only for the benefit of the student.

Slab base dimensions

The stanchion loads for this example result from dead plus imposed loading, with wind loading not taken into account. The base loads are axial and the slab base plan dimensions are determined directly from the base load and the allowable bearing pressure on the concrete foundation. The slab base thickness is derived from the following formula taken from BS449:

$$t = \sqrt{\left(\frac{3w}{\text{Pbct}}\left(A^2 - \frac{B^2}{4}\right)\right)}$$

where: t = slab thickness (mm); A = the greater projection of the plate beyond the stanchion (mm); B = the lesser projection of the plate beyond the stanchion (mm); w = the pressure on the underside of the base (N/mm^2); Pbct = the permissible bending stress in the steel (185 N/mm^2).

Calculations for slab base

For this purpose the stanchions are grouped into two classes, viz. ⑥, ⑦, ⑧, and the remainder. The allowable bearing pressure on the concrete foundation is assumed to be 5/N/mm^2. For 6, 7, 8 the maximum load = 1,792 kN. Area required

$$= \frac{1,792 \times 10^3}{5}$$
$$= 358,400 \text{ mm}^2$$
$$\therefore \text{ Sides} = \sqrt{358,400}$$
$$= 599, \text{ say } 600$$
$$w = \frac{1,792 \times 10^3}{600 \times 600}$$
$$= 4.98 \text{ N/mm}^2$$

Overhang dimensions $A = B = (600 - 254)/2$ = 173.

$$t = \sqrt{\left(\frac{3 \times 4.98}{185}\cdot\left(173^2 - \frac{173^2}{4}\right)\right)}$$
$$= 42, \text{ use } 45$$

Use 600 × 600 × 45.
For the remainder the maximum load = 753 kN. Area required

$$= \frac{753 \times 10^3}{5}$$
$$= 150,600 \text{ mm}^2$$

$$\therefore \text{ Sides} = \sqrt{150,600}$$
$$= 388, \text{ say } 400$$
$$w = \frac{753 \times 10^3}{400 \times 400}$$
$$= 4.7 \text{ N/mm}^2$$

Overhang dimensions $A = B = (400 - 203.2)/2$ = 98.4.

$$t = \sqrt{\left(\frac{3 \times 4.7}{185}\cdot\left(98.4^2 - \frac{98.4^2}{4}\right)\right)}$$
$$= 23.5 \text{ use } 25$$

Use 400 × 400 × 25

REQUISITION LISTS

Listing and allocation

The material required for the main structural members is listed on steel requisition sheets by the detailer, each item being given a reference mark. All members having the same serial size, mass per metre length and length are grouped together, and consequently have the same reference mark.

The material may be available ex-stock and allocated accordingly, or may have to be ordered either from steel stockholders or direct from the rolling mills. This information is added to the requisition sheet by office staff who deal with the ordering and allocation of the material. Typical lists are produced towards the end of this chapter (see Figs. 41 and 42).

Use of prints

Prints of the general arrangement and framing plans are produced before the steel requisition sheets are commenced.

Immediately particulars of each member or group of members are entered on the requisition sheets, coloured lines are drawn on the prints, along and covering the entire length of each listed member. This indicates clearly to all concerned that these members are listed and ensures that they will not be mistakenly entered a second time on the sheets.

The corresponding reference mark given to a member or group of members on the requisition sheet is then placed on each member on the prints, for cross reference purposes when material cutting lists are being prepared. These

CURPE CONSULTANTS

STEEL REQUISITION

		SHEET NO.	
REQD. DELIVERY DATE		No. OF SHEETS	JOB No. M2
		MADE BY	EST. No.
			DATE 9-6-80

Pitched Roof with Trusses

CLIENT:
DESCRIPTION OF JOB: PITCHED ROOF WITH TRUSSES

POSITION	REF. MARK	No. OFF	SECTION	WT.	LENGTH	No. OFF	LENGTH	REMARKS
Side Stans	1	8	254 x 146 UB	31	5025			
Gable Stans	2	4	152 x 152 UB	23	5300			
Do.	3	4	Do.	23	7000			
Do.	4	2	Do.	23	8300			
Purlins	5	12	100 x 65 x 7 L		8625			
Do.	6	12	Do.		4205			
Do.	7	12	Do.		8105			
Rafter bracing	8	16	90 x 90 x 6 L		2200			
Do.	9	2	Do.		4400			
Long. ties	10	6	80 x 80 x 6 L		4200			
Roof trusses								
Rafters	11	16	80 x 60 x 6 L		8310			
Side M.bil	12	16	75 x 50 x 6L		4400			
Centre M.bil	13	8	Do.		6200			
Internals	14	8	70 x 70 x 6 L		1800			
Do.	15	8	Do.		4400			
Do.	16	16	50 x 50 x 6L		1000			
Do.	17	16	Do.		2200			
Do.	18	4	Do.		3200			
M.tie bracing	19	4	40 x 40 x 6 L		5500			
Do.	20	4	Do.		5000			
Eaves ties	21	4	90 x 80 x 6L		6000			
Do.	22	10	127 x 76 I		4405			
Do.	23	8	Do.		8775			
Vert bracing	24	8	90 x 80 x 6 L		6000			
Side framing	25	4	90 x 90 x 6 L		8265			
Do.	26	4	Do.		4205			
Do.	27	4	Do.		8100			
Do.	28	4	Do.		9200			
Do.	29	1	Do.		8200			
Gable framing	30	1	203 x 76 L		9200			
Do.	31	2	Do.		2800			
Do.	32	2	90 x 90 x 6L		4005			
Do.	33	9	Do.		7525			
Do.	34	4	Do.		7755			
Do.	35	4	70 x 70 x 6 L		9300			

Fig. 42 (background). A steel requisition list for a pitched roof with trusses.

STEEL REQUISITION

Multi-Storey Building

CLIENT:
DESCRIPTION OF JOB: MULTI-STOREY BUILDING

POSITION	REF. MARK	No. OFF	SECTION	WT.	LENGTH	No. OFF	LENGTH	REMARK
Stanchions	1	3	254 x 254 UC	73	9155			
Do.	2	9	203 x 203 UC	46	8825			
Do.	3	1	Do.	46	9025			
Do.	4	4	Do.	46	7450			
Do.	5	3	Do.	46	7465			
Do.	6	6	203 x 133 UB	25	7450			
Roof Beams	7	4	254 x 144 UB	31	5722			
Do.	8	2	406 x 140 UB	39	5722			
Do.	9	4	Do.	39	5924			
	10	8	203 x 133 UB	25	4405			
	11	2	457 x 141 UB	47	7000			
Floor Beams	12	12	406 x 178 UB	60	5722			
	13	6	406 x 140 UB	44	5722			
	14	12	Do.	44	5924			
	15	24	254 x 144 UB	31	4405			
	16	6	610 x 229 UB	125	7000			

Fig. 41 (insert). A steel requisition list for a multi-storey building.

same prints can also be used at the detailing stage. When a member is detailed it can easily be distinguished from the remainder by drawing a line on the print in a similar fashion to that described for requisitioning, but in a different colour.

Lengths of members

The lengths of the members which are listed on the steel requisition sheets are generally only approximate lengths, overestimated to allow for eventual cutting to exact lengths. Some of the exact cutting lengths are established when shop detail drawings are completed and the remainder when template shop work is finalised.

Trusses and bracing

In the case of roof trusses the lengths of the members are scaled from the framing plans. Alternatively a rough detail layout of the truss, drawn to scale, will provide a more accurate means of assessment. An allowance of 50 mm is added to the estimated length of each member to give the lengths shown on the requisition sheets.

Lengths of vertical, main tie and rafter bracing members can be estimated in a similar fashion.

Purlins

For purlins an allowance of 25 mm is added to the estimated length, e.g. a purlin continuous over two spans, each of 4,000 mm, would be listed as 8,025 mm.

Stanchions

For stanchions the listed length is the calculated or estimated length plus an allowance of 50 mm.

Beams

For beams fitting between the flanges of stanchions the clear distance is calculated and an allowance of 25 mm is added to give the listed length.

For beams fitting between the webs of stanchions or beams an allowance of 25 mm is added to the distance, centre to centre, of stanchions or beams to give the listed length.

Splayed or bevelled ends

In the case of beams or stanchions with ends not cut square, an allowance of up to 150 mm may be given, e.g. for the vertical leg or stanchion of a portal frame or for a portal rafter.

General notes

Note that although there is a slight difference between the lengths of third floor beams marked (11), (61), (31), (71), (51), (81) and their corresponding beams at first and second floors, they have been grouped together on the steel requisition sheet produced for the multi-storey building example.

Note also that stanchions ⑥, ⑦ and ⑧ have a horizontal division plate 25 mm thick at the splice position (shown on the stanchion schedule) which accounts for the slight difference in the upper lengths of stanchions.

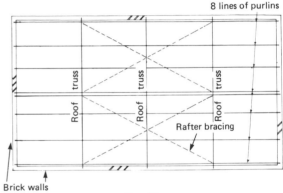

Fig. 43.

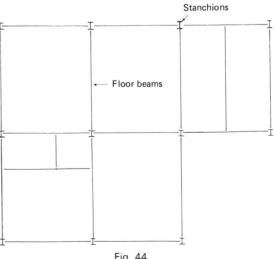

Fig. 44.

SELF-ASSESSMENT QUESTIONS

1. Supply erection marks for the purlins, roof trusses and rafter bracing shown on Fig. 43. Indicate the position of splices in the purlins.

2. Supply erection marks for the beams shown on the floor plan, Fig. 44, using: (*a*) the grid system; (*b*) any other standard system.

3. Prepare a steel requisition sheet for the main members of: (*a*) the flat roof construction in Fig. 34; (*b*) the pitched roof with portal frames shown in Fig. 37. Make out your own requisition sheet.

4. List the essential items to be included in preparing a stanchion schedule.

5. Explain the importance of placing beam section sizes, eccentricities, direction of floor spans and erection marks on steel framing plans.

CHAPTER FIVE

Stanchion Base Details

CHAPTER OBJECTIVES

After studying this chapter you should be able to:
• prepare foundation plans;
• prepare stanchion base details.

INTRODUCTION

When the general layout and framing plans are completed, the structural steelwork subcontractor prepares shop detail drawings for fabrication purposes, together with a foundation plan.

One of the earlier drawings required by the main contractor is the foundation plan. This drawing shows information needed when setting out the positions and levels of the stanchion foundations on the site and also when making provision for the fixing of the stanchions to the concrete.

FOUNDATION PLANS

Using the general arrangement and framing plans in conjunction with the architect's and/or consultant's drawings, the detailer prepares the foundation plan indicating:

(a) stanchion marks or grid lines;
(b) stanchion centre lines;
(c) layout dimensions, centre to centre, of stanchions;
(d) depth from floor level to top of foundations;
(e) holding down (HD) bolts;
(f) grout space.

As an illustration, a foundation plan for the multi-storey building example in Chapter Four is shown in Fig. 45. The dimension of 25 mm shown on the centre lines of stanchions ⑥ and ⑧ is determined from the details shown in Fig. 39 in Chapter Four.

HD bolts

Typical HD bolts are shown in the foundation plan. These are used to fix the steel stanchions to the concrete foundations. To allow for adjustment the bolts are cast in pockets formed in the foundation by means of tapered timber boxes, polystyrene sleeves or tubular sleeves which are removed when the foundations have been completed.

The bolts are noted on the drawing as CUP □ OX, representing cup head, square neck (at the washer plate), round shank, and hexagon nut respectively. A square washer plate (with a square hole) is provided for each HD bolt, thereby increasing the bolt's pull-out resistance from the concrete foundation.

The centres of the HD bolts are determined when the plan dimensions of the steel base plate are calculated. The projection length of the bolts above the top of the concrete foundation is established by adding together the dimensions shown on Fig. 46. The tolerance of 15 mm may be increased to 20 mm if desired.

Grout space

To allow for correction of inaccuracies in foundation levels a grouting space is introduced by inserting steel wedges or shims between the concrete and steel. This space and the space around the HD bolts are eventually filled in with cement grout when the steelwork has been lined and plumbed. A space of 25 mm is shown but this may be increased to 50 mm in certain cases.

STANCHION BASES

The loads at the base of a steel stanchion must

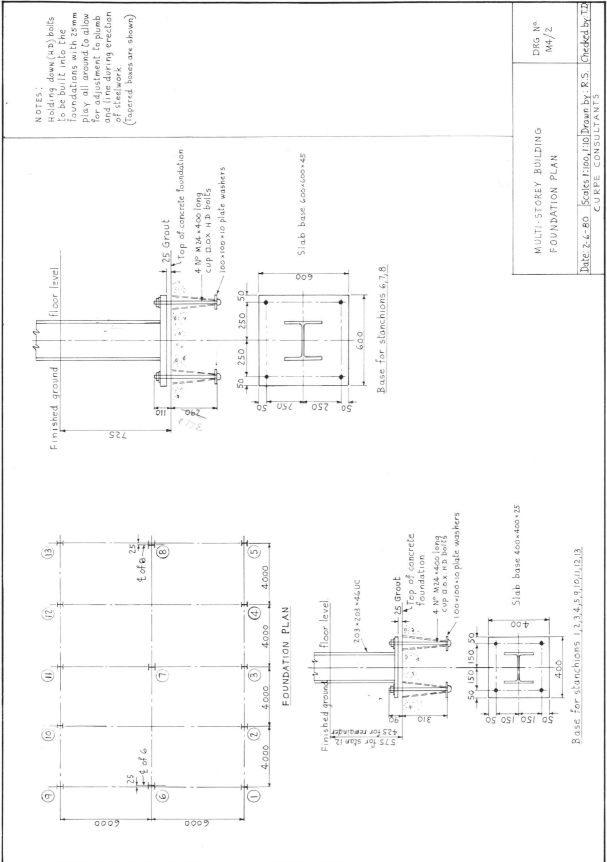

FOUNDATION PLAN

NOTES:
Holding down (HD) bolts
to be built into the
foundations with 25mm
play all around to allow
for adjustment to plumb
and line during erection
of steelwork.
(Tapered boxes are shown)

Base for stanchions 6,7,8

Slab base 600×600×45

25 Grout
Top of concrete foundation
4 Nº M24×400 long
cup Dox HD bolts
100×100×10 plate washers

floor level
Finished ground

Base for stanchions 1,2,3,4,5,9,10,11,12,13

Slab base 400×400×25

203×203×46UC

25 Grout
Top of concrete
foundation.
4 Nº M24×400 long
cup Dox HD bolts
100×100×10 plate washers

floor level
Finished ground

575 for stan 12
425 for remainder

MULTI-STOREY BUILDING
FOUNDATION PLAN

DRG Nº
M4/2

Date: 2-6-80 | Scales 1:100, 1:10 | Drawn by: R.S. | Checked by: T.D
CURPE CONSULTANTS

Fig. 45.

(a)

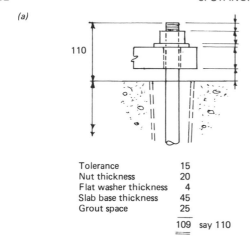

Tolerance	15
Nut thickness	20
Flat washer thickness	4
Slab base thickness	45
Grout space	25
	109 say 110

(b)

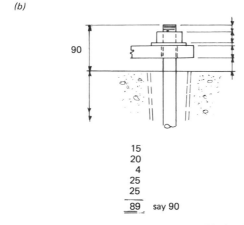

	15
	20
	4
	25
	25
	89 say 90

Fig. 46. *Establishing the projection length of bolts above top of concrete foundation.* (*a*) Projection of HD bolts for stanchions ⑥, ⑦, ⑧ in Fig. 45. (*b*) Projection of HD bolts for remainder of stanchions in Fig. 45.

be spread or distributed to the foundations by means of steel slab bases or stiffened/gusseted steel base plates, ensuring that the allowable bearing stress on the concrete is not exceeded. The slabs or plates are subjected to bending stresses and must be made thick enough to resist these stresses. Alternatively, a thinner plate stiffened by ribs or gussets may be designed. A thick base plate with no stiffeners requires less fabrication and is generally more economical.

Axial loads

Stanchion bases may be loaded either axially or eccentrically. For axially loaded bases the bearing pressure on the foundation is assumed to be uniformly distributed and the HD bolts are needed only to hold the stanchion securely in position.

Axial load plus bending

If a base is subjected to bending moments, in addition to axial load, the pressure on one edge of the base is greater than at the opposite edge.

If the moment is comparatively large and the vertical load small, then tension may occur at one side and HD bolts are necessary to resist the resultant uplift.

Axial load plus bending is outside the scope of this book, and will not be discussed any further.

Types of stanchion bases

The types to be considered are:

(*a*) slab;
(*b*) gusseted;
(*c*) pocket.

The details shown in Figs. 47–51 illustrate various types used in practice.

The following points will serve as an aid towards better understanding of the details.

(*a*) Holes for the HD bolts are drilled 4 mm larger in diameter than the diameter of the bolt (instead of the normal 2 mm) to facilitate the locating and fixing down of the stanchion base plate or slab base to the HD bolts in the concrete foundation.

(*b*) Minimum edge distance required to sheared or hand flame cut edges is 50 mm for a 30 mm diameter hole, and 38 mm for a 24 mm diameter hole. The dimensions shown on the details fulfil these requirements.

(*c*) "Shop" bolts are not normally dimensioned on detail drawings as they are positioned in the template shop or fabricating shop at standard cross centres in the flanges of I sections and at standard back marks on the angles used for flange and web cleats. All dimensions for shop bolts are shown on the details only for the benefit of the student.

(*d*) All bolt heads on the underside of the base

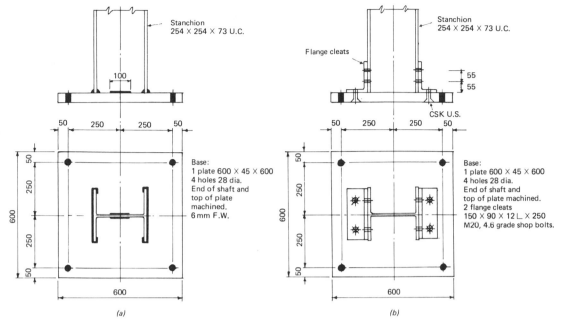

Fig. 47. *Alternative slab base types.* (*a*) Welded. (*b*) Bolted. (Scale 1 : 10.)

plates or slab bases are countersunk (CSK) to provide a flush surface.

(*e*) The top of the slab base or base plate and the end of the stanchion shaft may be machined for complete bearing. If so, instructions should be noted on the detail drawings.

(*f*) Different methods of representing and describing welds are illustrated.

(*g*) Maximum length of angle cleats in the web (web cleats) of stanchions is restricted to the specified depth between fillets (obtained from BCSA/CONSTRADO *Structural Steelwork Handbook*) for each particular section (refer also to particulars given at the end of Chapter Three, Tables 12–16). The lengths specified are satisfactory in this respect.

To accommodate bolts opposite one another in each leg of the web cleat and avoid the problem of overcrowding, a 90 × 90 × 10 L section is selected rather than an angle with short legs.

Slab bases
These are suitable for heavily loaded stanchions in multi-storey buildings and for single storey stanchions subjected to vertical loads, horizontal loads, and moments.

Example. The slab base for stanchions ⑥, ⑦ and ⑧ in Fig. 38 is taken as an example and Fig. 47 illustrates alternative methods of construction. These are by:

(*a*) welding;
(*b*) bolting.

The top of the slab and the end of the stanchion shaft are machined and therefore the welds or bolts are required only to retain the parts securely and permanently in place. Shop bolts are positioned to standard back marks and cross centres.

Fillet welds with return ends are indicated by thick lines and referred to as 6 mm FW.

Example. Figure 48(*a*) illustrates a typical slab base for a single storey stanchion subjected to vertical load, horizontal load, and moment, and Fig. 48(*b*) illustrates a small base suitable for a portal frame with pin or hinge bases. The stanchion and slab bases are not machined and welds are required to transmit loads and moments for Fig. 48(*a*) and loads only for Fig. 48(*b*).

In Fig. 48(*a*) a 6 mm fillet weld all round is

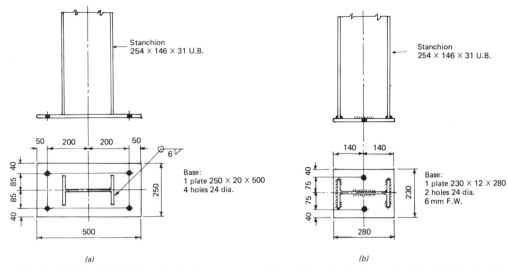

Fig. 48. (*a*) *Typical slab base for single storey stanchion subjected to horizontal and vertical load and moment.* (*b*) *Small base suitable for portal frame.* (Scale 1 : 10.)

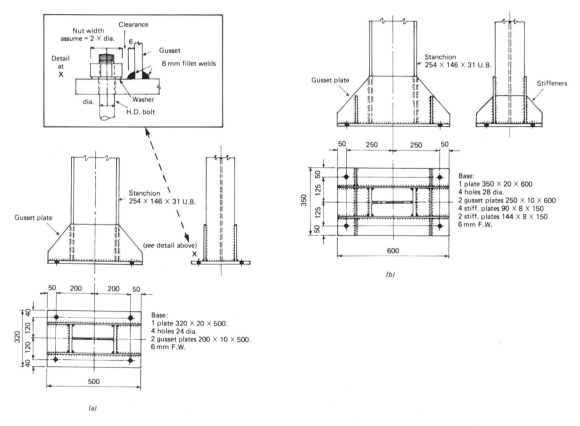

Fig. 49. (*a*) *A gusseted base.* (*b*) *A gusseted base with stiffeners.* (Scale 1 : 10.)

indicated by the appropriate symbol in conjunction with a reference line and arrow.

In Fig. 48(b) only holes for two HD bolts are shown and the 6 mm fillet welds are indicated by short 45° sloped lines.

Gusseted bases

Example. Figure 49(a) illustrates a gusseted base and Fig. 49(b) a gusseted base with the addition of stiffeners, both of welded construction. Generally a thinner plate may be used than for the equivalent slab base because of the additional support provided by gussets and stiffeners and the consequent reduction in the magnitude of the bending moments.

In both Figs. 49(a) and 49(b) the parts are not machined and the plates and welds are required to resist all applied forces and moments. The depth of the gusset plates is determined after the required length of welding is calculated.

It is important to arrange for sufficient room between the gusset welding to the base plate and the HD bolts to ensure that the nuts can be easily screwed on and tightened with a spanner without obstruction (*see* detail at X).

Example. Figure 50 illustrates gusseted bases of bolted construction, not machined for bearing.

In Fig. 50(a) bolts in the flange cleats and web cleats transmit the loads and moment to the base plate.

In Fig. 50(b) additional gusset plates are included to transmit loading and moment of greater magnitude to the base plate. Gusseted bases are suitable for heavily loaded stanchions or lightly loaded stanchions with substantial moments, e.g. a very high to eaves single storey stanchion subjected to significant wind moments. Bolts in the gussets, flange cleats and web cleats transmit the loads and moments to the base plate. The depth of the gusset plates is determined after the required number of bolts has been calculated.

Note that the dimension 175 in Fig. 50(a) is obtained from:

$$\frac{D}{2} + \text{BM} = \frac{251.5}{2} + 50$$
$$= 175.75, \text{ say } 175$$

(BM = back mark.)

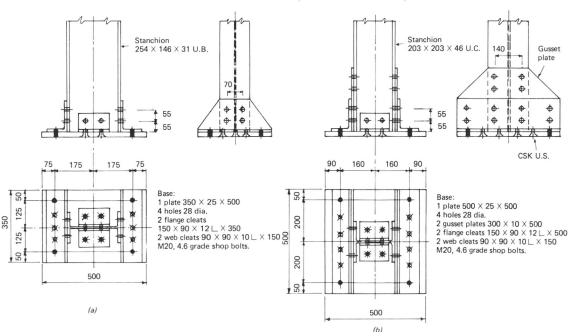

Fig. 50. *Gusseted bases of bolted construction.* (a) Bolts in flange and web cleats transmit loads and moment to base plate. (b) Additional gusset plates included to transmit loading and moment of greater magnitude to base plate. (Scale 1 : 10.)

The dimension 160 in Fig. 50(*b*) is similarly obtained from:

$$\frac{D}{2} + \text{Gusset} + \text{BM} = \frac{203.2}{2} + 10 + 50$$
$$= 161.6, \text{ say } 160$$

Pocket bases

These are not used as frequently as slab or gusseted bases, but are suitable for lightly loaded stanchions or for the base of portal frames designed with moment or fixed bases.

The stanchions are lowered into pockets formed in the concrete foundation. No base plate or HD bolts are necessary. Alternative methods for erection are illustrated in Fig. 51.

SELF-ASSESSMENT QUESTIONS

1. Explain the function of HD bolts and their washer plates. How is adjustment provided for the bolts in the concrete foundation?

2. To a scale of 1 : 10 draw two elevations and a plan for a gusseted base of bolted construction given the following particulars: stanchion size 254 × 254 × 73 UC (D × B = 254 × 254); base plate 500 × 25 × 500; gusset plates 225 × 10 × 500; flange cleats 150 × 90 × 12 L × 500; web cleats 90 × 90 × 10 L × 150; all shop bolts M20, 4.6 grade. Provide holes for No. 4, M24, HD bolts.

3. For the base detailed in Question 2, and using a scale of 1 : 10, show its connection to a concrete foundation given that the length of the HD bolts is 300 mm and the washer plates are 150 × 10 × 150. Dimension the projection and embedment length of the HD bolts.

4. A slab base for a stanchion of section 254 × 254 × 73 UC has the following particulars: slab 650 × 50 × 650; flange cleats 150 × 90 × 12 L × 250; all shop bolts M20, 4.6 grade. Draw two

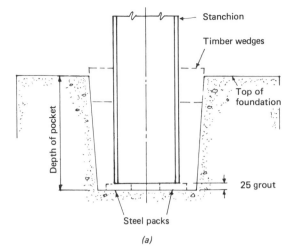

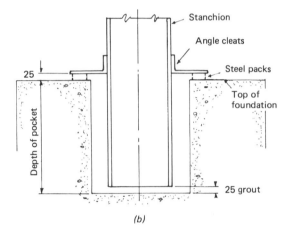

Fig. 51. *Alternative pocket type bases.* (Scale 1 : 10.)

elevations and a plan of the base to a scale of 1 : 10. Provide holes for No. 4, M24, HD bolts.

5. Explain the importance of a foundation plan to the: (*a*) general contractor; (*b*) steel erector.

CHAPTER SIX

Stanchion and Beam Connections

CHAPTER OBJECTIVES

After studying this chapter you should be able to:
* prepare stanchion details;
* prepare beam details;
* prepare stanchion and beam connections;
* prepare material requisition lists for connections;
* prepare material cutting lists;
* prepare bolt lists.

INTRODUCTION

Now that the foundation plan, stanchion base details and material requisition lists for the steel members have been finalised, preparation of the shop detail drawings can be undertaken by the detailer. When shop detail drawings are completed, material requisition lists for the connections, i.e. cleats, plates, cuttings, are made out and sufficient material allocated accordingly, thus ensuring its availability when fabrication commences.

Material cutting lists are made out for the steel members, followed by separate cutting lists for the connections.

Lengths of shop bolts (for assembly purposes) and site bolts (for erection purposes) are determined and listed separately.

DETAILING OF STANCHIONS AND BEAMS

In this chapter connectors are mentioned, types of connections described and illustrated and complete details produced for a stanchion and several beams. In practice stanchions are detailed first and when completed the beam details are commenced.

Connectors

The principal connectors used for structural steel connections are:

(*a*) fillet welds;
(*b*) butt welds;
(*c*) black bolts, grade 4.6;
(*d*) friction grip bolts.

In this chapter black bolts, grade 4.6, have been used for bolted connections. A diameter size of 20 mm is commonly used by fabricators, and this is therefore generally adopted (unless reduced to suit flange widths or leg lengths of I or angle sections respectively).

Beam to stanchion connections

These are generally of the following types:

(*a*) top and bottom (or seating) cleats (*see* Fig. 52);
(*b*) web and bottom (or seating) cleats (*see* Fig. 53);
(*c*) welded end plates (*see* Fig. 54);
(*d*) web cleats (*see* Fig. 54);
(*e*) eccentric connection (*see* Fig. 56);
(*f*) cap plates (*see* Fig. 57).

Top and bottom cleats (*see* Fig. 52)
The bottom cleats are shop bolted to the stanchion and the top cleats are shop bolted to the beam.

The end reactions are assumed to be resisted entirely by the bolts in the vertical legs of the bottom cleats, the top cleats only steadying the top of the beams.

The end clearance to the top cleats for beams (B1) and (B2) is determined from:

$$\frac{D}{2} + 2 = \frac{206}{2} + 2$$
$$= 105$$

The end clearance to the top cleats for beams (B3) is determined from:

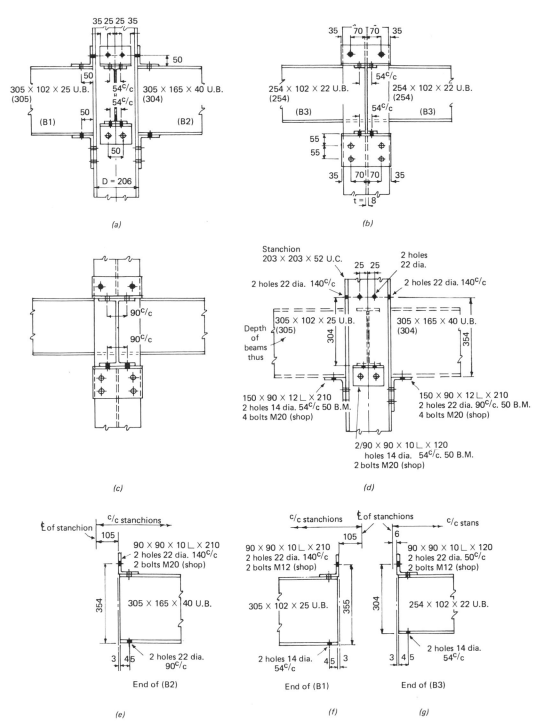

Fig. 52. *Top and bottom cleats.* Elevations (*a*), (*b*) and (*c*) show typical connections of this type. (*d*) shows the method adopted when detailing a stanchion. (*e*), (*f*) and (*g*) show details at the ends of beams (B1), (B2) and (B3), corresponding to the stanchion detail at (*d*). (Scale 1 : 10.)

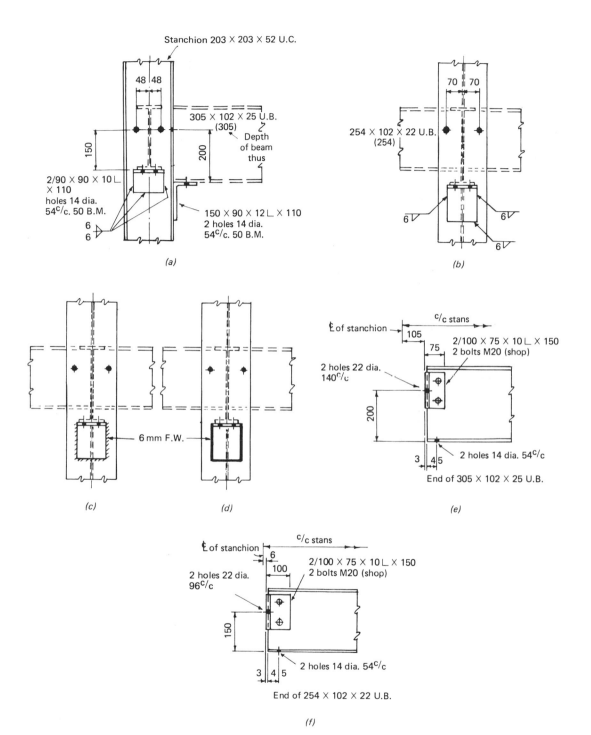

Stanchion 203 × 203 × 52 U.C.

48 | 48

150

200

305 × 102 × 25 U.B.
(305)
Depth
of beam
thus

2/90 × 90 × 10 L
× 110
holes 14 dia.
54^c/c. 50 B.M.

6
6

150 × 90 × 12 L × 110
2 holes 14 dia.
54^c/c. 50 B.M.

(a)

70 | 70

254 × 102 × 22 U.B.
(254)

6

6

6

(b)

6 mm F.W.

(c)

(d)

^{c}/c stans

₵ of stanchion

105

75

2/100 × 75 × 10 L × 150
2 bolts M20 (shop)

2 holes 22 dia.
140^c/c

200

3 | 4 | 5

2 holes 14 dia. 54^c/c

End of 305 × 102 × 25 U.B.

(e)

^{c}/c stans

₵ of stanchion

6

100

2/100 × 75 × 10 L × 150
2 bolts M20 (shop)

2 holes 22 dia.
96^c/c

150

3 | 4 | 5

2 holes 14 dia. 54^c/c

End of 254 × 102 × 22 U.B.

(f)

Fig. 53. *Web and bottom cleats*. (Scale 1 : 10.)

TABLE 17. SHEARING AND BEARING VALUES OF BLACK BOLTS (kN)

| Dia. of bolt shank (mm) | No. of bolts | Shearing value at 80 N/mm² | | Bearing value at 250 N/mm² for end distance of 2.0 × dia of bolt | | | | | | | | | | |
| | | | | Thickness in mm of plate passed through | | | | | | | | | | |
		Single	Double	5	6	7	8	9	10	12	15	18	20	22
M12	1	9	18	15	18	21								
	2	18	36	30	36	42								
	3	27	54	45	54	63								
	4	36	72	60	72	84								
	5	45	90	75	90	105								
	6	54	109	90	108	126								
	7	63	127	105	126	147								
	8	72	145	120	144	168								
	9	81	163	135	162	189								
	10	90	181	150	180	210								
M16	1	16	32	20	24	28	32	36						
	2	32	64	40	48	46	64	72						
	3	48	97	60	72	84	96	108						
	4	64	129	80	96	112	128	144						
	5	80	161	100	120	140	160	180						
	6	97	193	120	144	168	192	216						
	7	113	225	140	168	196	224	252						
	8	129	257	160	192	224	256	288						
	9	145	290	180	216	252	288	324						
	10	161	322	200	240	280	320	360						
M20	1	25	50	25	30	35	40	45	50	60				
	2	50	101	50	60	70	80	90	100	120				
	3	75	151	75	90	105	120	135	150	180				
	4	101	201	100	120	140	160	180	200	240				
	5	126	251	125	150	175	200	225	250	300				
	6	151	302	150	180	210	240	270	300	360				
	7	176	352	175	210	245	280	315	350	420				
	8	201	402	200	240	280	320	360	400	480				
	9	226	452	225	270	315	360	405	450	540				
	10	251	503	250	300	350	400	450	500	600				

Black bolts to be to BS4190, grade 4.6, passing through grades 43 and 50 material.

$$C = \frac{t}{2} + 2$$
$$= \frac{8}{2} + 2$$
$$= 6$$

Note that the top cleats project 3 mm beyond the end of the beams.

Bolts. Bolts resisting the end reaction from the beams may fail either in shear or bearing. Black bolts, grade 4.6 and 20 mm diameter, are selected for the connections. The allowable load for this bolt is determined in single shear, double shear and bearing as follows:

Allowable load in single shear = Allowable stress × Cross section area

$$= 80 \times \frac{\pi \times 20^2}{4} \times \frac{1}{10^3}$$
$$= 25 \text{ kN}$$

Allowable load in double shear
$$= 2 \left(80 \times \frac{\pi \times 20^2}{4} \right) \frac{1}{10^3}$$
$$= 50 \text{ kN}$$

Allowable load in bearing per mm thickness =
Allowable stress × Bolt diameter × Thickness
$$= \frac{250 \times 20 \times 1}{10^3}$$
$$= 5 \text{ kN}$$

Shearing and bearing values for black bolts, grade 4.6, are tabulated and are given in Table 17. The allowable stresses are obtained from BS499.

Ends of (B1) and (B2). Bolts may fail in single shear or bearing. Bearing thickness is the lesser of the cleat thickness and the stanchion flange thickness, i.e. 12 mm or 12.5 mm.

∴ Allowable load in bearing is $5 \times 12 = 60$ kN

Allowable load per bolt is the lesser of the shear and bearing values.

∴ Allowable load per bolt = 25 kN
Allowable load for four bolts = 25 × 4
= 100 kN

Ends of (B3). Bolts may fail in double shear or bearing. Bearing thickness is the lesser of the combined thicknesses of the two cleats or the stanchion web thickness, i.e. (2 × 10) mm or 8 mm.

Allowable load in bearing is $5 \times 8 = 40$ kN
∴ Allowable load per bolt = 40 kN
Allowable load for two bolts = 40 × 2
= 80 kN

i.e. 40 kN allowable reaction per beam.

Web and bottom cleats (see Fig. 53)
The bottom cleats are welded (or may be shop bolted) to the stanchion, with the web cleats being shop bolted to the beam. The end reactions are assumed to be resisted entirely by the welds shown on the three sides of the bottom cleats. Three methods of illustrating and describing the welds are shown at (*a*), (*b*), (*c*), and (*d*).

Welds. The allowable load per mm run for a fillet weld is:

Allowable stress × Throat thickness × 1 mm

For a 6 mm fillet weld the allowable load is:
$$\frac{115 \times 0.7 \times 6}{10^3} = 0.48 \text{ kN per mm run}$$

Allowable load = Effective length of weld
× Allowable load per mm
= (150 + 150 + 110 − 12) × 0.48
= 191 kN

The effective length is the total length less (2 × 6) mm for end craters. Elevations (*e*) and (*f*) show the details at the ends of the beams.

The end clearance to the web cleats for the 305 × 102 × 25 UB is:
$$\frac{D}{2} + 2 = \frac{206}{2} + 2$$
$$= 105$$

The end clearance to the web cleats for the 254 × 102 × 22 UB is:
$$C = \frac{t}{2} + 2$$
$$= \frac{8}{2} + 2$$
$$= 6$$

Note that the web cleats project 3 mm beyond the end of the beam.

Welded end plates (see Fig. 54)
Elevations (*a*) and (*b*) illustrate the stanchions, with (*d*) and (*e*) the corresponding details at the ends of the beams. The dimension 104 is obtained from:
$$\frac{D}{2} + 2 = \frac{203.2}{2} + 2$$
$$= 103.6, \text{ say } 104$$

The depth of weld available at (*d*) for strength calculations is taken as the depth *d* between the fillets, i.e. 359 mm.

Allowable load for the welds at (*d*) is:
$$0.48 \times 2 \times 359 = 334 \text{ kN}$$

Allowable load for the six bolts is:
$$25 \times 6 = 150 \text{ kN}$$

Allowable load is controlled by the bolts and is 150 kN.

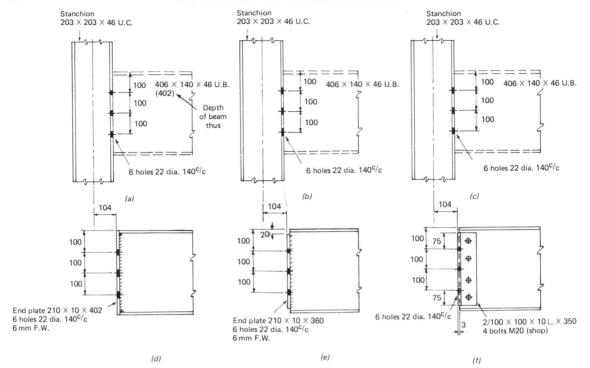

Fig. 54. *Welded end plates*. (Scale 1 : 10.)

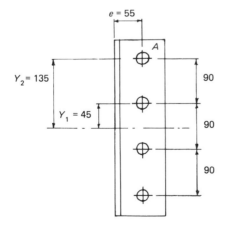

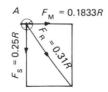

Fig. 55. *Web cleat*. (Scale 1 : 5.)

Web cleats (*see* Fig. 54)

Elevation (*f*) illustrates the use of web cleats, with (*c*) the corresponding stanchion. The dimension 104 is obtained from $(D/2) + 2$, as for welded end plates. Cleats project 3 mm beyond the end of the beam. Allowable load for the connection is the lesser of the allowable load for the four bolts in the connected legs and the six bolts in the outstanding legs of the cleats.

Figure 55 shows the dimensions for the spacing of the bolts in the connected leg and the forces on bolt A.

Bolts in a connected leg are subjected to direct force, and the force due to the moment of magnitude $R \times e$, where R is the end reaction in kN and e is the eccentricity in mm.

The resultant force on the extreme bolt can be obtained by calculation. This value must not exceed the allowable load for the bolt.

If N is the number of bolts and a is the area of one bolt (mm²):

$$\text{Direct force } F_s = \frac{R}{N} = \frac{R}{4}$$

$$= 0.25R \text{ kN}$$

Connection for beams mark (92), (102), (112) etc. at 1st and 2nd floor levels.

Direct force $F_S = \dfrac{R}{N} = \dfrac{(62 + 62)}{6} = 20.66$

Moment $= 124 \times 30$

Modulus of group $= \Sigma r^2 = 4(r_1^2) + 2(r_2^2)$

$= 4(71.06^2) + 2(45^2) = 24250$

Force due to moment $F_M = \dfrac{Rer_1}{\Sigma r^2}$

$F_M = \dfrac{124 \times 30 \times 71.06}{24250} = 10.9$

$F_R = \sqrt{F_S^2 + F_M^2 + 2F_S.F_M \cos\Theta}$

$\cos\Theta = \dfrac{45}{71.06} = 0.63 \quad F_R = 28.8$

Allowable load double shear $= 50$ kN
Allowable load bearing $= 5 \times 7.3 = 36.5$
$\therefore$ Satisfactory

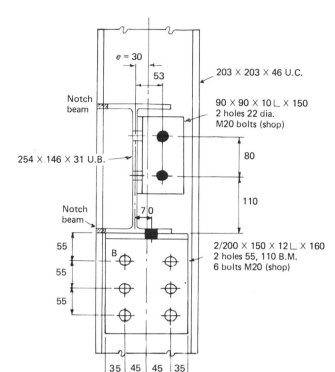

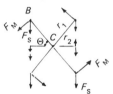

Forces in bolts.
C is the centre of rotation.

Fig. 56. *Eccentric connection (multi-storey building example).* (Scale 1 : 5.)

Force due to bending = Bending stress × Area of bolt

Moment = Reaction × Eccentricity
$= 55R$ kN mm

Moment of inertia of bolt group $I = a(2Y_1^2 + 2Y_2^2)$ mm⁴

$Z = \dfrac{I}{Y} = a\dfrac{[2(45)^2 + 2(135)^2]}{135}$

$= 300a$ mm³

$F_M =$ Force due to moment

$= \dfrac{M}{Z} \times a$

$F_M = \dfrac{55R \times a}{300a}$

$= 0.1833R$

Resultant force F_R on bolt A $= \sqrt{(F_S^2 + F_M^2)}$

$F_R = \sqrt{((0.25R)^2 + (0.1833)^2)}$

$F_R = 0.31R$

Allowable load per bolt is the lesser of double shear value and the bearing value.

Allowable load in double shear = 50 kN.

Allowable load in bearing for a beam web thickness of 6.9 mm = 5 × 6.9 = 34.5 kN.

$\therefore$ Allowable load is 34.5 kN for one bolt
$0.31R = 34.5$
$R = 111.3$ kN

Bolts in outstanding legs must be checked for single shear and bearing.

Allowable load in single shear = 25 kN
Allowable load in bearing = 5 × 10 = 50 kN

Allowable load for six bolts = 25 × 6
$= 150$ kN

$\therefore$ Allowable load is controlled by lesser of 111.3 or 150 and is 111.3 kN.

Eccentric connection (see Fig. 56)
In this example, taken from the multi-storey

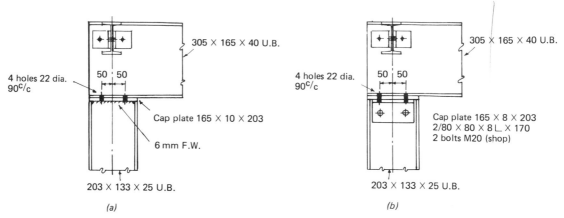

Fig. 57. *Cap plates.* (*a*) Welded directly to the stanchion. (*b*) Bolted to cleats which are shop bolted to stanchions. (Scale 1 : 10).

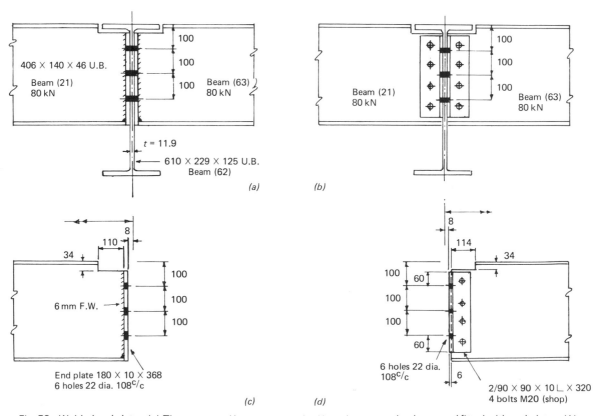

Fig. 58. *Welded end plates.* (*a*) The supported beams are notched into the supporting beam and fitted with end plates. (*b*) As for (*a*) but with a web cleat. (*c*) The ends of the beams. (*d*) As for (*b*). (Scale 1 : 10.)

building described in Chapter Four, the six bolts must be checked for the direct force and the force due to the moment caused by 30 mm eccentricity.

Cap plates (see Fig. 57)
Cap plates are either welded directly to the stanchion as at (*a*), or bolted down to cleats which are shop bolted to the stanchion as shown at (*b*).

Beam to beam connections
These are generally of the following types:

(*a*) Welded end plates (*see* Fig. 58);
(*b*) Web cleats (*see* Fig. 58);
(*c*) Web and bottom cleats.

Welded end plates (see Fig. 58)
Figure 58 shows the connection between floor beams (21), (63), and (62), for the multi-storey building example in Fig. 38.

In (*a*) the supported beams (21) and (63) are notched into the supporting beam (62) and fitted with end plates. Elevation (*c*) illustrates the ends of the beam (21) and (63) with end clearance and notching dimensions indicated.

The end clearance of 8 mm is obtained from:

$$\frac{t}{2} + 2 = \frac{11.9}{2} + 2$$
$$= 7.95, \text{ say 8 mm}$$

The allowable load for the welds is:

$$0.48 \times 2 \times 359 = 344 \text{ kN/beam}$$

The bolts are in double shear. The bearing thickness is the lesser of the combined thickness of the two cleats, or the web thickness of the $610 \times 229 \times 125$ UB, i.e. $(10 + 10)$ or 11.9 mm.

Allowable load in double shear = 50 kN

Allowable load in bearing = 5×11.9
$$= 59.5 \text{ kN}$$

Allowable load for six bolts is:

$$50 \times 6 = 300 \text{ kN}$$

Allowable load for the connection is controlled by the lesser of the weld and bolt loads, i.e. 300 kN. This is satisfactory since the allowable load is greater than the actual load of 160 kN.

Web cleats (see Fig. 58)
In (*b*) and (*d*), web cleats are shown as an alternative to the welded end plates. Note that the cleats project 6 mm beyond the end of the beam.

Web and bottom cleats
In this type, the bolts in the vertical leg of the bottom cleat resist the end reaction. Web cleats steady the beams.

Stanchion details
All stanchions that are identical in every respect are grouped together, and one stanchion is completely detailed. The number required is noted accordingly.

Complete details for stanchion S7 of the multi-storey building example are shown in Fig. 59.

The details are drawn to a scale of 1 : 10, but the vertical heights from roof to floor and floor to floor levels are not to scale.

All connections, i.e. plates and cleats, are given assembly marks (A), (B), (C), (D), etc. All plates or cleats which are identical in every respect are given the same mark. At the splice where the top section is of different width to the bottom section, a division plate (E) and plate packs (G) are necessary. The length of cleats in the webs of the stanchions is restricted to the depth *d* between the fillets, i.e. 200.2 mm for the $254 \times 254 \times 73$ UC section and 160.8 mm for the $203 \times 203 \times 46$ UC section. Note that in the horizontal legs of cleats (B), holes have been reduced to 18 mm diameter because of insufficient edge distance for the connecting beam.

Note that the $610 \times 229 \times 125$ UB floor beams and the $457 \times 191 \times 98$ UB roof beams, coming into the web of the top stanchion, are to be connected by means of plates shop welded to the ends of the beams and site bolted to the stanchion web, i.e. end plate type. The flanges of these beams will of course require notching into the stanchion.

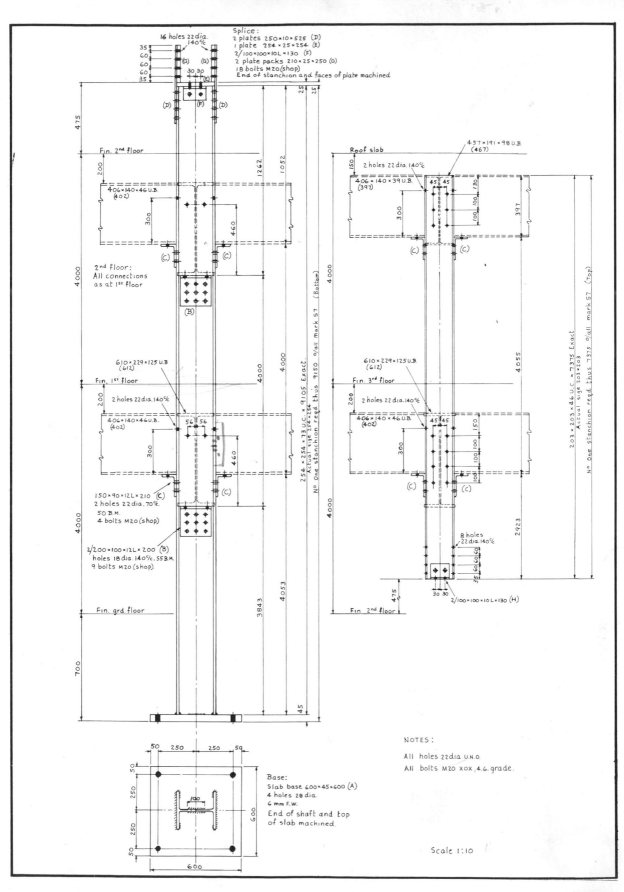

Fig. 59.

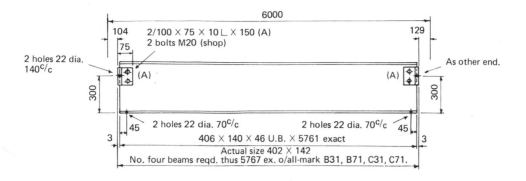

6000

104 2/100 × 75 × 10 L × 150 (A)
     2 bolts M20 (shop)   129

2 holes 22 dia.
140 c/c   As other end.
75

300   (A)   (A)   300

45 2 holes 22 dia. 70 c/c   2 holes 22 dia. 70 c/c 45
3   3
       406 × 140 × 46 U.B. × 5761 exact
              Actual size 402 × 142
No. four beams reqd. thus 5767 ex. o/all-mark B31, B71, C31, C71.

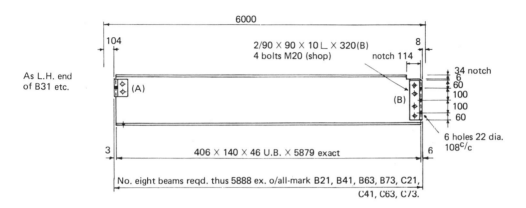

6000

104   2/90 × 90 × 10 L × 320(B)   8
                       4 bolts M20 (shop)   notch 114

As L.H. end   34 notch
of B31 etc.   6
              (A)   60
                                              (B)   100
                                                         100
                                                         60

                                                    6 holes 22 dia.
3   108 c/c
       406 × 140 × 46 U.B. × 5879 exact   6

No. eight beams reqd. thus 5888 ex. o/all-mark B21, B41, B63, B73, C21,
                                              C41, C63, C73.

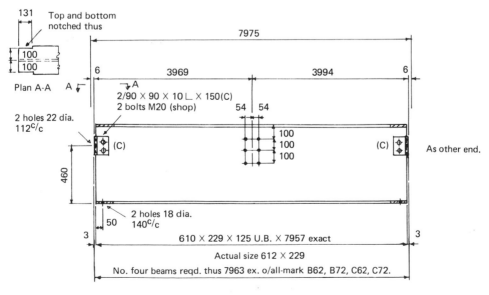

131 Top and bottom
     notched thus   7975

100
100   6   3969   3994   6

Plan A-A   A   A

2 holes 22 dia.   2/90 × 90 × 10 L × 150(C)
112 c/c   2 bolts M20 (shop)   54 54

              (C)   100   (C)   As other end.
460   100
                                           100

          2 holes 18 dia.
50   140 c/c
3   3
       610 × 229 × 125 U.B. × 7957 exact
              Actual size 612 × 229
No. four beams reqd. thus 7963 ex. o/all-mark B62, B72, C62, C72.

Fig. 60. *Typical beams selected from Fig. 38.* (Scale 1 : 20.)

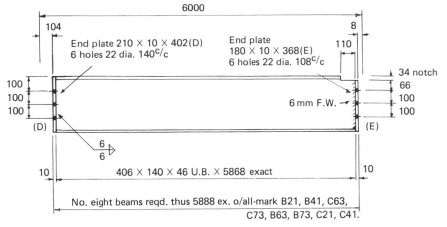

Fig. 61. *Detail of beam* (B21) *from Fig. 60.* (Scale 1 : 20.)

Bolts for cleat (*B*)

To illustrate the method of determining the number of bolts required in, for example, cleats (B), the following calculations are produced.

$$\text{Reaction} = 160 + 160$$
$$= 320 \text{ kN}$$

Allowable load in double shear = 50 kN

Allowable load in bearing = 5 × stanchion web thickness

$$= 5 \times 8.6$$
$$= 43 \text{ kN}$$
$$\text{Number of bolts required} = \frac{320}{43}$$
$$= 7.44$$

Use nine as shown.

Base

The base detail corresponds to that shown on the foundation plan.

Beam details

All beams identical in every respect are grouped together and one beam is completely detailed. The number required is noted accordingly.

Typical selected beams from the multi-storey building example are detailed in Fig. 60.

The details in this instance are drawn to a scale of 1 : 20, but should preferably be done to a larger scale of 1 : 10. The length of the beams, however, are not drawn to scale. The top dimension above the beam is the centre to centre distance of stanchions or beams. End clearances to cleats or welded end plates are first established, with cleats

projecting beyond the end of the beams as shown.

All plates and cleats are given assembly marks (A), (B), (C), etc. All plates or cleats which are

STEEL REQUISITION

CLIENT:

DESCRIPTION OF JOB   MULTI - STOREY BUILDING

POSITION	REF. MARK	No. OFF	SECTION	WT.	LENGTH		No. OF
CONNECTIONS TO STANCHION S 7 (Fig. 59)							
PLATES							
Slab base	17	1	600×45 Plate		600	Ex	
Splice	18	2	250×10 Plate		525	Ex	
Do.	19	1	254×10 Plate		254	Ex	
Packs	20	2	210 × 25 Plate		250	Ex	
CLEATS							
Seating	21	4	200×100×12 L		200	Ex	
Do.	22	8	150 × 90 ×12 L		210	Ex	
Splice	23	2	100×100×10 L		130	Ex	
Do.	24	2	100×100 ×10 L		130	Ex	
CONNECTIONS TO BEAMS (Fig. 60)							
CLEATS							
End Web	25	32	100 × 75 ×10 L		150	Ex	
Do.	26	16	90 × 90 × 10 L		320	Ex	
Do.	27	16	Do.		150	Ex	

Fig. 62. *Steel requisition list for connections in Figs. 59 and 60.*

CURPE CONSULTANTS

WORKS ORDER	Made By R.A.S.	Date 20-6-1980	DRG. NO. Fig. 59 / Fig. 60	NAME MULTI-STOREY BUILDING	ADDRESS	SHEET NO.

MARKS	No. OFF	SECTION	LENGTH	PAINTED	Ref. Mark
STAN. 57	1	254 × 254 × 73 U.C.	9105	Ex.	1
Do.	1	203 × 203 × 46 U.C.	7375	Ex.	5
Beams B31, B71, C31, C71	4	406 × 140 × 46 U.B.	5761	Ex.	13
Beams B31, B41, B63, B73, C21, C41, C63, C73	8	406 × 140 × 46 U.B.	5879	Ex.	14
Beams B62, B72, C62, C72	4	610 × 229 × 125 U.B.	7157	Ex.	16

CURPE CONSULTANTS

WORKS ORDER	Made By R.A.S.	Date 20.6.80	DRG. NO. Fig. 59 / Fig. 60	NAME MUL	ADDRESS	SHEET NO. 1.

MARKS	No. OFF	SECTION	LENGTH	PAINTED	Ref. Mark
CONNECTIONS TO STANCHION S7 (Fig. 59)					
Plates					
Slab base (A)	1	600 × 45 Plate	600	Ex.	17
Seating (D)	2	250 × 10 Plate	525	Ex.	18
Splice (D)	2	254 × 10 Plate	254	Ex.	19
Do. (E)	1	210 × 25 Plate	250	Ex.	20
Packs (G)	2		200	Ex.	21
CLEATS					
Seating (B)	4	200 × 100 × 12 L	200	Ex.	22
Do. (C)	8	150 × 90 × 12 L	210	Ex.	23
Splice (F)	2	100 × 100 × 10 L	130	Ex.	24
Do. (H)	2	100 × 100 × 10 L	130	Ex.	
CONNECTIONS TO BEAMS (Fig. 60)					
CLEATS					
end web (A)	32	100 × 75 × 10 L	150	Ex.	25
Do. (B)	16	90 × 90 × 10 L	320	Ex.	26
Do. (C)	16	Do.	150	Ex.	27

Fig. 63 (background). Cutting list for stanchion (ST) and the selected beams from Figs. 59 and 60.

Fig. 64. (insert) Cutting list for the relevant connections in Figs. 59 and 60.

identical in every respect are given the same mark. Figure 61 is drawn solely to show beams B21, etc., detailed with welded end plates as an alternative method of construction to the details shown for beams B21, etc., on Fig. 60. Fillet welds are illustrated and described in two different ways, either way being satisfactory.

REQUISITION LISTS FOR CONNECTIONS

From the multi-storey building example, a selected stanchion mark S7 has been detailed on Fig. 59, and also typical selected beams on Fig. 60. Requisition lists for the connections on Figs. 59 and 60 are shown on Fig. 62.

The reference marks are continued from those shown on the previous sheet No. 1, Fig. 41, which ended at 16. The required material is allocated from stock or ordered if necessary.

MATERIAL CUTTING LISTS

Cutting lists are prepared for stanchion (S7) and the selected beams (*see* Fig. 63) and for their relevant connections (*see* Fig. 64). Exact lengths of members are marked Ex to indicate that these can be measured and cut to this length in the fabricating shop.

Reference marks shown must correspond with those given originally to the relevant members on Fig. 41. Reference marks shown on sheet No. 2 must correspond with those given to the relevant connections on Fig. 62.

BOLT LISTS

The types and diameters of the bolts are specified on the detail drawings. Eventually the lengths have to be determined, and for this purpose two sets of typical bolt lists are prepared, one for shop bolts (used at assembly stage), the other for site bolts (used at erection stage), although for convenience both are shown on Fig. 65. When all bolts for the steelwork are listed a summary sheet is made out for site bolts, grouping together bolts of the same type, diameter and length. Bags of bolts corresponding to these requirements are despatched to the site for the erection of the steelwork.

The individual thicknesses passed through are added together to obtain the grip. The length of the bolt is the grip plus an allowance for a washer, depth of nut and sufficient projection beyond the nut. Examples are:

(*a*) M20 black bolt: length = grip + 3 + 16 + say 5;
(*b*) M16 black bolt: length = grip + 3 + 13 + say 5;
(*c*) M20 HSFG bolt: length = grip + 30;
(*d*) M16 HSFG bolt: length = grip + 26;

These lengths obtained are rounded off to the nearest 5 mm above and are based on the use of one flat round washer.

For taper flanges of sections it is necessary to use taper washers and a special allowance for thickness has to be given. The typical bolt list made out is for stanchion S7 (from base level to splice level); the top length from splice level to roof level is left as an exercise for the student.

SELF-ASSESSMENT QUESTIONS

1. Referring to Fig. 34, Scheme 2, prepare to a scale of 1:10 a shop detail drawing for stanchion S2. Include: (*a*) two elevations; (*b*) plan on the base plate. The base plate, 250 × 20 × 480, is to be provided with four holes for M20, HD bolts and is to be welded to the stanchion with 6 mm fillet welds all round. Use M20, grade 4.6 black bolts for the beam connections.

2. Referring to Fig. 34, Scheme 2, prepare to a scale of 1:10 a shop detail drawing for the beams B2 and B5. Tie beams B5 are to be provided with two cleats 150 × 100 × 10L × 80 each end for connection to the main beams B2. Use M20, grade 4.6 black bolts.

3. Referring to Fig. 38, prepare to a scale of 1:10 detailed connections for the following: (*a*) first floor beam connections to stanchion S3; (*b*) roof beam connections to stanchion S3.

4. Referring to Fig. 38, prepare to a scale of 1:10 a detailed sketch of the connection between beams 21, 63 and 62 at roof level.

5. Figure 65 shows shop bolts and site bolts for stanchion S7, base level to splice level. Prepare separate lists for shop and site bolts for stanchion S7 from splice level to roof level. Refer to Fig. 59.

6. Prepare material requisition lists for the stanchion and beams detailed in answer to Questions 1 and 2.

7. Prepare material requisition lists for the

CURPE CONSULTANTS

Site .... Name MULTI - STOREY BUILDING

See Drawing No......Fig. 59......................... Contract No................ Date 20-6-1980

No.	Type	Thicknesses Through				Grip	Add	Length	Dia	Head	Neck	Nut	Position	Wash-ers
	SHOP BOLTS FOR STANCHION					57	(base level to splice level)							
16	4·6 grade black	12	14			26	24	50	20	x	0	x	cleat (c) to stan. flange	FRW
18	Do.	12	12	9		33	24	60	20	x	0	x	cleat (B) to stan. web	FRW
16	Do.	10	14			24	24	50	20	x	0	x	splice plate (CD) to stan. flange	FRW
2	Do.	10	10	9		29	24	55	20	x	0	x	splice cleats (f) to stan. web	FRW
	SITE BOLTS FOR STANCHION					57	(base level to splice level)							
8	4·6 grade black	12	11			23	24	50	20	x	0	x	Beams 31, 71 to cleat (c)	FRW
8	Do.	10	14			24	24	50	20	x	0	x	Beams 31, 71 to stan. flange	FRW
4	Do.	10	10	9		29	24	55	20	x	0	x	Beams 62, 72, to stan. web	FRW
8	Do.	12	20			32	21	55	16	x	0	x	Beams 62, 72 to cleat (B)	FRW

F.R.W. — Flat round washers.
X.O.X. — Hexagon round hexagon.

Fig. 65. *Bolt list for Fig. 59.*

connections for stanchions and beams detailed in answer to Questions 1 and 2.

8. Prepare complete details for stanchion S3. Refer to Fig. 38.

9. Prepare cutting lists for the: (*a*) stanchions; (*b*) beams; (*c*) stanchion connections; (*d*) beam connections; detailed in answer to Questions 1 and 2.

FURTHER READING

Structural Steelwork—Design of Components
Structural Steelwork—Fabrication Manual on Connections
(All published by BCSA.)

CHAPTER SEVEN

Reinforced Concrete in General

CHAPTER OBJECTIVES

After studying this chapter you should be able to:
- appreciate the basic reasons for reinforcing concrete with steel;
- understand the general rules relating to this reinforcement.

INTRODUCTION

In Chapters One and Two we sought to introduce the subject of structural drawings by dealing with the reasons why drawings exist and their methods of production. In Chapters Three to Six the subject of detailing various structural steelwork elements was studied. Now we must deal with reinforced concrete.

There are a number of points which apply to reinforced concrete members generally and since these lay the basis for much of the detailing shown in subsequent chapters it is essential that you are familiar with these recommendations. They include the reasons for having reinforcement in concrete, the need for cover to the reinforcement, bar sizes available and their areas, and maximum and minimum areas of reinforcement in slabs, beams and columns, including links.

These rules are contained mostly in CP110, where we also find recommendations relating to spacing, grouping and curtailment of bars. These are dealt with as they arise in the chapters which deal with the relevant concrete element.

DESIGN CRITERIA

The design of reinforced concrete structures has increased in complexity since the first attempt at a code of practice in 1934. New codes of practice were introduced and revised between 1948 and 1965. At that time there existed three booklets, all of modest length—CP114, CP115 and CP116, covering reinforced, prestressed and precast concrete structures.

In October 1964 committees were established to consider a revision of the three codes mentioned above in the light of the work of the European Concrete Committee (CEB), with particular consideration to incorporating the design theory of limit states. As this proceeded it became clear that all the work ought to be gathered together and a new unified code established to cover the three ranges of concrete work mentioned above, i.e. reinforcement, prestressed and precast concrete, mainly in relation to buildings (there are other codes which govern bridge works). This eventually led to the publication in 1972 of CP110 *The Structural Use of Concrete*, a document consisting of three formidable volumes.

It is the recommendations within this code of practice which are used throughout the concrete sections of this book.

DESIGN PRINCIPLES

It is not the purpose of this book to pursue the methods of design and analysis laid down in CP110, but it is nevertheless true that only when the structural detailer has grasped the broad design principles will he be competent in his work.

For example, why is the reinforcement in a cantilever heavier in one face than the other? Why is the main reinforcement in a beam heavier in the bottom of the span than at the top, and yet at the support this is reversed? Why in a wall is the horizontal reinforcement sometimes placed on the outside of the vertical reinforcement? These and other questions can be answered by someone with a grasp of the simple and basic design principles set out below.

One of the fundamental properties of concrete is that it is weak when in tension and yet comparatively strong when in compression. The ratio is approximately $10:1$ in favour of compression.

Now consider the simply supported beam which deflects under load, as shown in Fig. 66.

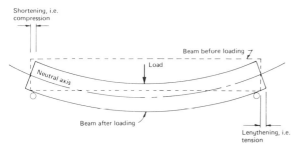

Fig. 66. *Simply supported beam deflected under load.*

It will be appreciated that the bottom of the beam is being stretched, i.e. in tension, while the top is being pushed, i.e. in compression. This beam will crack and thus fail when the ultimate tensile stress is reached and yet it will not have reached its ultimate compressive stress. The beam is therefore incapable of utilising its compressive strength because of the weakness in tension.

If steel, which is strong in tension and will readily bond to the concrete, is introduced in the tension zone it will resist the tensile stresses while the concrete will resist the compressive stresses, thus producing a stronger beam, i.e. the beam is reinforced. Hence the name reinforced concrete.

In very simple terms it can be said that the structural engineer's job is to calculate where the concrete is likely to fail and to design sufficient reinforcement to prevent failure taking place. A detailer, without having an understanding of the complexities of design, can appreciate where the main reinforcement should be positioned by simply considering the deflected form of the structure.

Deflected forms

Figure 67 shows the deflected form of a number

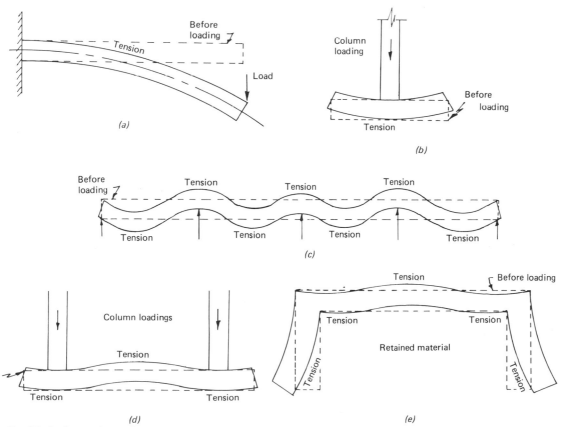

Fig. 67. *Deflected form of a number of concrete structures.* (*a*) Cantilevered structure. (*b*) Isolated foundation. (*c*) Continuous structure. (*d*) Combined foundation. (*e*) Plan of retaining walls.

of concrete structures and the position of the tensile stresses is indicated on each. Since concrete is comparatively weak in tension, as stated above, the main reinforcement would be positioned in the areas of tensile stress.

You will note in the case of the wall, which is comparatively short in plan with stiff return walls, that it is possible for such a wall to span horizontally rather than vertically and this results in the main reinforcement being placed horizontally on the outside face.

COVER TO THE REINFORCEMENT

The cover to the reinforcement is the thickness of concrete between the reinforcement and the edge of the concrete section (*see* Fig. 68).

There are at least three reasons why this dimension is important.

(*a*) It prevents the penetration of moisture into the section. If water were to reach the reinforcement it would cause corrosion and eventually the concrete would flake off around the bar, i.e. the concrete would spall off.

(*b*) It prevents fire reaching the reinforcement quickly. When the fire does reach the reinforcement this causes expansion, breaking the bond

between the concrete and steel and thus hastening failure.

(*c*) It prevents the reinforcement slipping within the concrete section. If the cover was too small the concrete and reinforcement could not act together, since the steel would slip and the bond would be broken.

The first two of these reasons is for durability and the third for design. There is, however, a fourth consideration which also relates to design and a brief and simple explanation will underline the importance of clearly and correctly stating the cover on a drawing.

When designing a reinforced concrete member, one of the most important dimensions in an engineer's calculations is the "effective depth" of the section. This is the distance from the face of the concrete in compression to the centre of the reinforcement, as shown in Fig. 69.

This effective depth is based on the cover given to the reinforcement and on the sizes of the bars in the section. In his calculations the engineer uses the value of the effective depth squared, i.e. d^2. Thus the cover to the reinforcement takes on an increased significance since a slight change will considerably affect the design, e.g. if the engineer uses an effective depth of 700 mm in a particular calculation and the detailer reduces

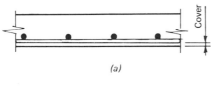

(a)

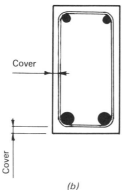

(b)

Fig. 68. *The cover to the reinforcement.* (*a*) Section through slab. (*b*) Section through beam.

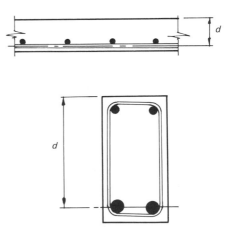

Fig. 69. *Effective depth of the section.*

TABLE 18. COVER FOR VARIOUS EXPOSURE CONDITIONS

Condition of exposure	Nominal cover (mm) for concrete grade				
	20	25	30	40	50 and over
Mild, e.g. completely protected against weather or aggressive conditions, except for brief period of exposure to normal weather conditions during construction	25	20	15	15	15
Moderate, e.g. sheltered from severe rain and against freezing whilst saturated with water; buried concrete and concrete continuously under water	—	40	30	25	20
Severe, e.g. exposed to driving rain, alternate wetting and drying and to freezing whilst wet; subject to heavy condensation or corrosive fumes	—	50	40	30	25
Very severe, e.g. exposed to sea water or moorland water and with abrasion	—	—	—	60	50
Subject to salt used for de-icing	—	—	50*	40*	25

* Only applicable if the concrete has entrained air.

the cover by 25 mm, this changes the effective depth to 675 mm, i.e. 3.57 per cent, but it changes the calculations not by 3.57 per cent but by the ratio of 700^2 to 675^2, i.e. 7.02 per cent.

Since the strength of the beam is related to the effective depth it will be appreciated why it is usual to place the main reinforcement on the outside and the secondary reinforcement on the inside, i.e. to give a larger effective depth for the same size of section.

The cover for various exposure conditions is given in Table 19 of CP110, and this is shown in Table 18. These recommendations are satisfactory where water is the eroding element; however, where there is a risk of exposure to fire, calcium chloride, or other aggressive atmospheres, or where there is a special surface treatment, e.g. bush hammering, the dimensions given will have to be increased.

In any case the nominal cover should always be at least equal to the size of the bar, and in the case of bundles of three or more bars should be equal to the size of a single bar of equivalent area. Table 18 gives the minimum nominal cover to all reinforcement including the links.

BAR SIZES AND AREAS

The final result of an engineer's calculations for a structural concrete element is an area of reinforcement which, when incorporated with the section designed, will produce an element which will safely withstand the applied loads. This area of reinforcement has to be given as a number of bars of the chosen diameter. The bar sizes which are commonly available are 6 mm, 8 mm, 10 mm, 12 mm, 16 mm, 20 mm, 25 mm, 32 mm and 40 mm diameter.

To assist the engineer in his choice of bars and spacings, standard steel area charts, as shown in Tables 19 and 20 have been prepared.

Alternatively the reinforcement may consist of flat sheets of reinforcement which are formed into a fabric. These have bars tack welded at right angles to each other and is known as fabric or mesh. This is most useful when dealing with ground floor slabs and walls which have a fairly simple and repetitive plan layout. Where this is complex or where numerous openings are required it is better to detail bars.

The British Standard preferred types of mesh

TABLE 19. STEEL AREA CHART FOR NUMBER OF BARS

Bar diameter (mm)	Areas in mm² for numbers of bars									
	1	*2*	*3*	*4*	*5*	*6*	*7*	*8*	*9*	*10*
6	28	57	85	113	142	170	198	226	255	283
8	50	101	151	201	252	302	352	402	453	503
10	79	157	236	314	392	471	550	628	707	785
12	113	226	339	452	566	679	792	905	1,020	1,130
16	201	402	603	804	1,010	1,210	1,410	1,610	1,810	2,010
20	314	628	943	1,260	1,570	1,890	2,200	2,510	2,830	3,140
25	491	982	1,470	1,960	2,450	2,950	3,440	3,930	4,420	4,910
32	804	1,610	2,410	3,220	4,020	4,830	5,630	6,430	7,240	8,040
40	1,260	2,510	3,770	5,030	6,280	7,540	8,800	10,100	11,300	12,600

TABLE 20. STEEL AREA CHART FOR SPACING OF BARS

Bar diameter (mm)	Areas in mm² for spacings in mm								
	50	*75*	*100*	*125*	*150*	*175*	*200*	*250*	*300*
6	566	377	283	226	189	162	142	113	94
8	1,010	671	503	402	335	287	252	201	168
10	1,570	1,050	785	628	523	449	393	314	262
12	2,260	1,510	1,130	905	745	646	566	452	377
16	4,020	2,680	2,010	1,610	1,340	1,150	1,010	804	670
20	6,280	4,190	3,140	2,510	2,090	1,800	1,570	1,260	1,050
25	9,820	6,550	4,910	3,930	3,270	2,810	2,450	1,960	1,640
32	16,100	10,700	8,040	6,430	5,360	4,600	4,020	3,220	2,680
40	25,100	16,800	12,600	10,100	8,380	7,180	6,280	5,030	4,190

are shown in Table 21. Others are indicated in Chapter Two, Fig. 13, as outlined by BS4466.

Example 1
A reinforced concrete column has been designed and the amount of main reinforcement required is 1,918 mm².

From Table 19 it can be seen that four 25 mm diameter bars give 1,960 mm², which is satisfactory.

Example 2
A reinforced concrete slab has been designed and the amount of main reinforcement required is 726 mm²/m.

From Table 20 it can be seen that 12 mm diameter bars at 150 mm centres give 745 mm²/m, which is satisfactory.

RULES RELATING TO AREAS OF REINFORCEMENT

Sections 3.11.4 and 3.11.5 of CP110 give the minimum and maximum areas of main, secondary and linkage reinforcement. These are now considered in relation to slabs, beams and columns.

Reinforced concrete slabs

The minimum area of main reinforcement
The minimum area of main reinforcement in tension in a slab should not be less than 0.15 per cent $b_t d$ when using high yield reinforcement or 0.25 per cent $b_t d$ when using mild steel reinforcement, where b_t is the breadth of the section and d is the effective depth.

TABLE 21. WIREWELD METRIC FABRIC—BRITISH STANDARD PREFERRED TYPES

	British Standard reference	Mesh size Nominal pitch of wires		Size of wires		Cross sectional area per metre width		Nominal mass per square metre (kg)
		Main (mm)	Cross (mm)	Main (mm)	Cross (mm)	Main (mm)	Cross (mm)	
Square mesh fabric:	A 393	200	200	10	10	393	393	6.16
	A 252	200	200	8	8	252	252	3.95
	A 193	200	200	7	7	193	193	3.02
	A 142	200	200	6	6	142	142	2.22
	A 98	200	200	5	5	98	98	1.54
Structural fabric:	B 1,131	100	200	12	8	1,131	252	10.9
	B 785	100	200	10	8	785	252	8.14
	B 503	100	200	8	8	503	252	5.93
	B 385	100	200	7	7	385	193	4.53
	B 283	100	200	6	7	283	193	3.73
	B 196	100	200	5	7	196	193	3.05
Long mesh fabric:	C 785	100	400	10	6	785	70.8	6.72
	C 636	100	400	9	6	636	70.8	5.55
	C 503	100	400	8	5	503	49.0	4.34
	C 385	100	400	7	5	385	49.0	3.41
	C 283	100	400	6	5	283	49.0	2.61

Sheet size and availability: square and long meshes are normally available from stock in sheets of 4.8 m × 2.4 m.

Example 3
Calculate the minimum area of main tension reinforcement in a slab 175 mm thick. Assume the main reinforcement is 12 mm diameter bars of mild steel with 20 mm cover.

For a 1 m width of slab:

Area of reinforcement = 0.25 per cent $b_t d$

where $b_t = 1,000$ mm and $d = 175 - 20 - (12/2)$
$$= 149 \text{ mm.}$$

∴ Area of reinforcement $= \dfrac{0.25 \times 1,000 \times 149}{100}$
$$= 372.5 \text{ mm}^2/\text{m}$$

From Table 20 12 mm bars at 300 mm centre to centre give 377 mm²/m, which is satisfactory.

The minimum area of secondary reinforcement
The minimum area of secondary reinforcement in a solid slab, expressed as a percentage of the gross cross section, should not be less than 0.12 per cent of high yield reinforcement or, alternatively, not less than 0.15 per cent of mild steel

reinforcement. In either case the distance between bars should not exceed five times the effective depth of the slab.

Example 4
Calculate the spacing of 10 mm mild steel bars required to provide secondary reinforcement for the slab in Example 3 above

Area = 0.15 per cent $b_t h$
$$= 0.15 \times \frac{1,000}{100} \times 175$$
$$= 262.5 \text{ mm}^2/\text{m}$$

From Table 20 10 mm bars at 300 mm centre to centre give 262 mm²/m, which can be considered as satisfactory.

Reinforced concrete beams

The minimum area of main reinforcement
The minimum area of main reinforcement in tension in a beam should not be less than 0.15

per cent $b_t d$ when using high yield reinforcement or 0.25 per cent $b_t d$ when using mild steel reinforcement. This is exactly the same as the requirement for slabs, and b_t and d are as defined for slabs.

Example 5
Calculate the minimum area of main tension reinforcement in a beam 350 mm wide and 600 mm deep. Assume the main reinforcement is 16 mm diameter bars of high yield reinforcement and the links are 8 mm diameter bars with 20 mm cover.

$$\text{Area} = 0.15 \text{ per cent } b_t d$$
$$d = 600 - 20 - 8 - \left(\frac{16}{2}\right)$$
$$= 564 \text{ mm}$$
$$\text{Area} = \frac{0.15 \times 350 \times 564}{100}$$
$$= 296.1 \text{ mm}^2$$

From Table 19 two 16 mm diameter bars give 402 mm², which is satisfactory.

The maximum area of main reinforcement
In a beam neither the area of tension reinforcement or the area of compression reinforcement should exceed 4 per cent of the gross cross-sectional area of the concrete. Where the percentage of reinforcement is approaching the maximum, special attention must be given to lap positions to ensure that the concrete can be properly placed and compacted.

Example 6
Calculate the maximum amount of reinforcement permissible in either tension or compression in the beam in Example 5 above.

$$\text{Area} = \frac{4}{100} \times 350 \times 600$$
$$= 8,400 \text{ mm}^2$$

From Table 19 this is equivalent to ten 32 mm diameter bars, which gives 8,040 mm².

The minimum area of links
Where the shear stress in a beam is greater than half the permissible value, links must be provided. These may be designed from Equation (9) in CP110, but in any case the minimum amount of links provided throughout the span should be such that for links of high yield reinforcement:

$$\frac{A_{sv}}{S_v} = 0.0012 b_t$$

or for links of mild steel:

$$\frac{A_{sv}}{S_v} = 0.002 b_t$$

where A_{sv} is the cross-sectional area of the two legs of a link; S_v is the spacing of the link; and b_t is the breadth of the beam at the level of the tension reinforcement.

Also the links should enclose all tension reinforcement, and their spacing either longitudinally or laterally should not exceed 0.75 times the effective depth.

Example 7
Calculate the minimum amount of shear reinforcement for the beam in Example 5 above.

$$A_{sv} \text{ for 8 mm links} = 101 \text{ mm}^2$$
$$\therefore S_v = \frac{101}{0.0012 \times 350}$$
$$= 240.48 \text{ mm}$$

say 225 mm centre to centre.
This is less than 0.75 times the effective depth, i.e. 423 mm, and is therefore satisfactory. Use 8 mm links at 225 mm centre to centre.

Reinforced concrete columns

The minimum amount of reinforcement
The minimum number of longitudinal bars provided in a column should be four in rectangular columns and six in circular columns, and their size should not be less than 12 mm diameter. Except for lightly loaded columns the total cross-sectional area of these bars should not normally be less than 1 per cent of the cross section of the column.

Example 8
Calculate the minimum amount of main reinforcement permissible in a concrete column having cross-sectional dimensions of 300 mm × 400 mm.

$$\text{Area} = \frac{1 \times 300 \times 400}{100}$$
$$= 1{,}200 \text{ mm}^2$$

From Table 19, four 20 mm diameter bars give 1,260 mm², which is satisfactory.

The maximum amount of reinforcement in a column

The maximum amount of reinforcement in a column is based on a constructional consideration, i.e. whether it is cast horizontally or vertically. The percentage of longitudinal reinforcement should not exceed 6 per cent for vertically cast columns and 8 per cent for horizontally cast columns, except at laps where in both types of column the percentage may be 10 per cent.

Example 9

Calculate the maximum amount of main reinforcement permissible in the column in Example 8 when:

(*a*) cast vertically; and
(*b*) cast horizontally.

(*a*)
$$\text{Area} = \frac{6 \times 300 \times 400}{100}$$
$$= 7{,}200 \text{ mm}^2$$

From Table 19 nine 32 mm diameter bars give 7,240 mm², or six 40 mm diameter bars give 7,540 mm², either of which is satisfactory.

(*b*)
$$\text{Area} = \frac{8 \times 300 \times 400}{100}$$
$$= 10{,}800 \text{ mm}^2$$

From Table 19 nine 40 mm diameter bars give 11,300 mm², which is satisfactory.

The minimum area of links

The minimum area of links in a column is given in relation to their size and spacing:

(*a*) at least one quarter the size of the largest compression bar;
(*b*) maximum spacing is twelve times the size of the smallest compression bar.

Example 10

Calculate the size and spacing of the links required for the column whose cross sectional details are shown in Fig. 70.

$$\begin{aligned} \text{Size} &= 0.25 \times 40 \\ &= 10 \text{ mm} \\ \text{Spacing} &= 12 \times 20 \\ &= 240 \text{ mm} \end{aligned}$$

Therefore use 10 mm diameter links at 240 mm centre to centre.

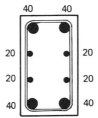

Fig. 70. *Cross section of column, bar diameters shown in mm.*

SELF-ASSESSMENT QUESTIONS

1. Indicate where you would expect the main reinforcement in the concrete members shown in Fig. 71.

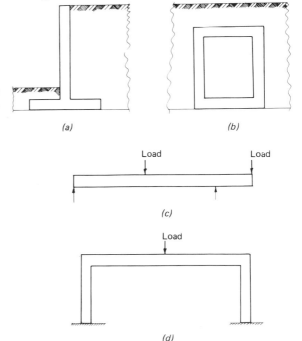

Fig. 71. (*a*) *Retaining wall.* (*b*) *Box culvert.* (*c*) *Beam.* (*d*) *Portal frame.*

2. A reinforced concrete foundation is to be constructed in grade 30 concrete. What would be the minimum cover?

3. A reinforced concrete slab is 200 mm thick and it requires the minimum amount of main reinforcement. If the slab is inside a building and the bars are 12 mm diameter of high yield reinforcement, what would be the spacing of the bars?

4. A beam is 300 mm wide and 550 mm deep. What would be the spacing of 10 mm diameter links of mild steel to give the minimum amount of shear reinforcement?

CHAPTER EIGHT

General Arrangement of a Reinforced Concrete Frame

CHAPTER OBJECTIVES

After studying this chapter you should be able to:
* prepare a complete set of general arrangement drawings for a reinforced concrete frame building.

FUNCTION

In Chapter One we saw that general arrangement drawings were produced showing the layout, typical floor and roof plans and sections through the building, all setting out clearly and unambiguously the proposed structural scheme. Where the job incorporates any features not revealed on the large sections, small part sections ought to be drawn to highlight these.

These drawings are used by the majority of the design team to enable preliminary design work to proceed. They also give the tendering contractor an easily assimilated impression of the overall structure without having to study the full set of working drawings. You will appreciate, therefore, that the general arrangements (GAs) are a very valuable asset to the members of the design and construction teams and must be produced with great care in order to ensure accurate information.

FORMAT

The format of the general arrangement drawings should comply with the recommendations previously outlined in Chapter Two. The basic and fundamental principles which must be grasped are those which relate to grid lines. These are used extensively to produce a comprehensive reference system for the complete structure.

FEATURES

It is not easy to give a foolproof list of all the information that should appear on general arrangement drawings, since each job usually has its own special features. However experience shows that the following information is usually given and is indeed essential:

(a) the layout of the structure in relation to the site;

(b) the size and materials of all elements of the structure;

(c) the concrete strength and maximum aggregate size used;

(d) the type of reinforcement used;

(e) the loading which the floors and roof have been designed to withstand.

Other information which may be given depending on the job is as follows:

(f) the means of access to the site and any limitations on movement within or around the site, i.e. site boundaries;

(g) any known obstructions such as underground services, overhead cables;

(h) existing and proposed ground levels;

(i) proposed drain and sewer runs;

(j) in some cases a method of construction could be suggested.

With reference to (a), the setting out would be incorporated on the general arrangement drawing for a simple job. Where this is more complex two individual drawings may be necessary. If the site is in a rural setting the grid system chosen must be referenced to the Ordnance Survey grid, enabling co-ordinates to be computed and plotted. In a built-up area the structure may be positioned by referring to existing adjacent buildings. Often the architect will locate the building and decide on the grid system leaving the structural engineer to work within an established pattern.

Points (c), (d) and (e), although listed on the general arrangement drawings, should also appear on the detailed drawings.

Items (*f*), (*g*), (*h*) and (*i*) are usually detailed on the architect's drawings in the case of building structures. However for heavy civil engineering jobs they would appear on the engineer's drawings.

In the case of (*j*) it is not common to outline the constructional procedure where it is traditional. Where this is not the case it is important to highlight any special features. This allows the contractor to appreciate the problems and thus produce a more realistic tender price. Where the stability of the structure depends on a particular sequence of construction it is important, from a safety aspect, for this to be clearly stated prior to the tendering stage.

Figures 72–77 seek to illustrate the layout, plans and sections which would be fairly typical of the work a structural detailer would have to produce.

SELF-ASSESSMENT QUESTIONS

1. Discuss the importance of a general arrangement drawing.

2. List the features of a general arrangement drawing, giving reasons why each is important.

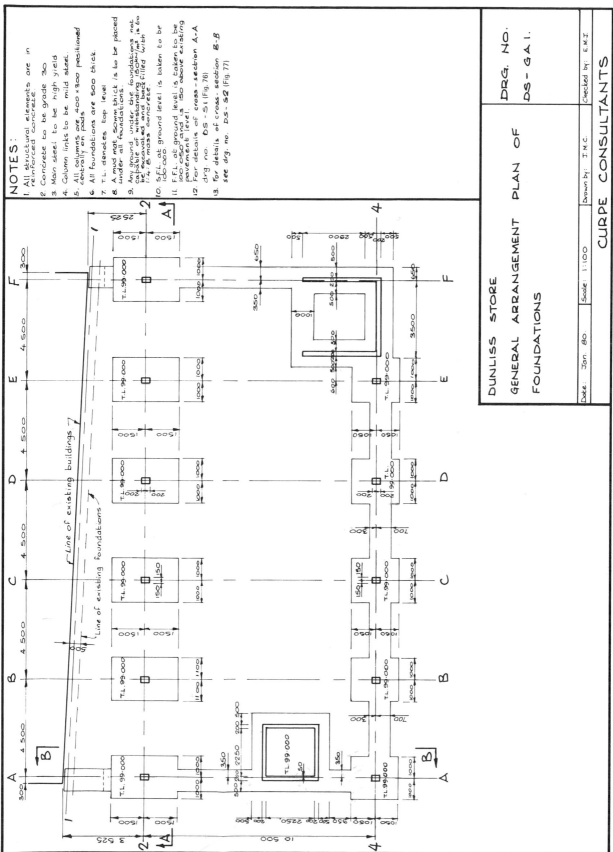

NOTES:

1. All structural elements are in reinforced concrete.
2. Concrete to be grade 30.
3. Main steel to be high yield.
4. Column links to be mild steel.
5. All columns are 400 x 300 positioned centrally on pads.
6. All foundations are 500 thick.
7. T.L. denotes top level.
8. A mud mat 50mm thick is to be placed under all foundations.
9. Any ground under the foundations not capable of withstanding 150kN/m² is to be excavated and backfilled with 1:4:8 mass concrete.
10. S.F.L. at ground level is taken to be 100·000.
11. F.F.L. at ground level is taken to be 100·050 and is 150 above existing pavement level.
12. For details of cross-section A-A drg. no. DS-S1 (Fig.76)
13. For details of cross-section B-B see drg. no. DS-S2 (Fig.77)

DUNLISS STORE
GENERAL ARRANGEMENT PLAN OF FOUNDATIONS

Date: Jan. 80.	Scale: 1:100	Drawn by: I.M.C.	Checked by: E.M.I

DRG. NO.
DS - GA1.

CURPE CONSULTANTS

Fig. 72.

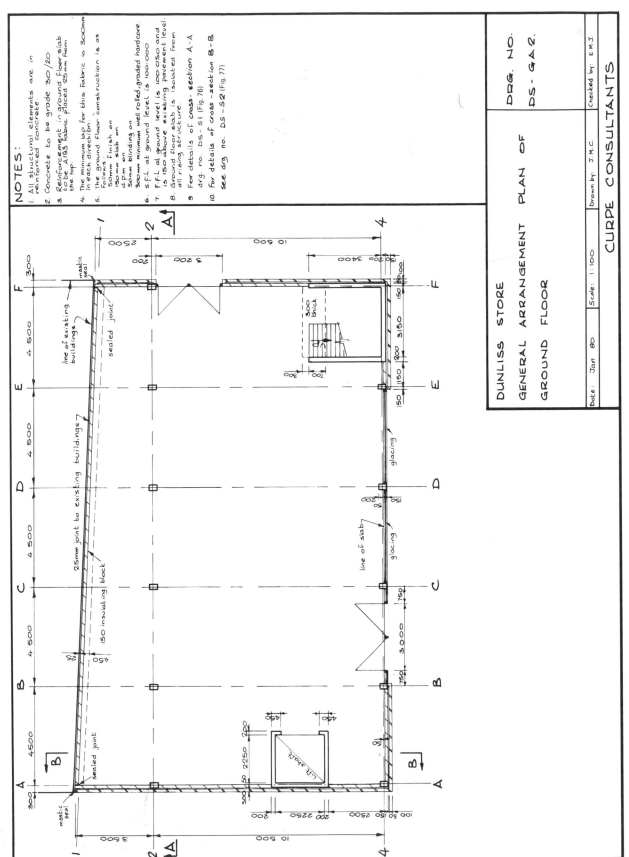

NOTES:

1. All structural elements are in reinforced concrete.

2. Concrete to be grade 30/20

3. Reinforcement in ground floor slab to be A193 fabric placed 25mm from the top.

4. The minimum lap for this fabric is 300mm in each direction.

5. The ground floor construction is as follows:
   50mm finish on
   150mm slab on
   d.p.m on
   50mm blinding on
   300mm minimum well rolled graded hardcore.

6. S.F.L. at ground level is 100·000

7. F.F.L. at ground level is 100·050 and is 150 above existing pavement level.

8. Ground floor slab is isolated from all rising structure.

9. For details of cross-section A-A drg. no. DS-S1 (Fig.76)

10. For details of cross-section B-B see drg no. DS-S2 (Fig.77)

DUNLISS STORE				DRG. NO.
GENERAL ARRANGEMENT PLAN OF				DS-GA2.
GROUND FLOOR				
Date: Jan 80	Scale: 1:100	Drawn by: J.M.C	Checked by: E.M.J	
CURPE CONSULTANTS				

Fig. 73.

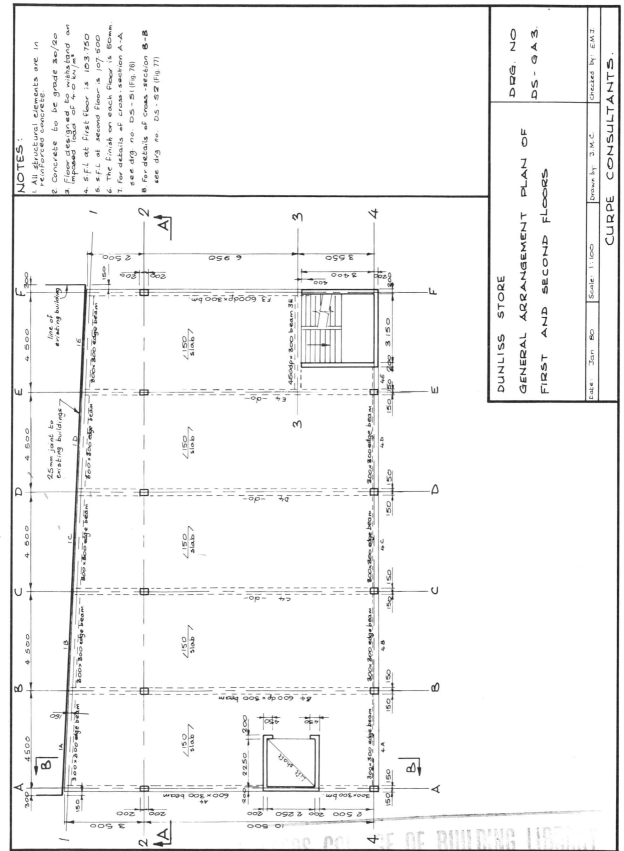

NOTES:

1. All structural elements are in reinforced concrete.
2. Concrete to be grade 30/20
3. Floor designed to withstand an imposed load of 4.0 kN/m²
4. s.f.l at first floor is 103.750
5. s.f.l at second floor is 107.500
6. The finish on each floor is 50mm.
7. For details of cross-section A-A see drg no. DS-S1 (Fig. 76)
8. For details of cross-section B-B see drg no. DS-S2 (Fig. 77)

DUNLISS STORE					DRG. NO
GENERAL ARRANGEMENT PLAN OF					DS-GA3
FIRST AND SECOND FLOORS					
Date: Jan 80	Scale: 1:100	Drawn by: J.M.C.	Checked by: E.M.J.

CURPE CONSULTANTS.

Fig. 74.

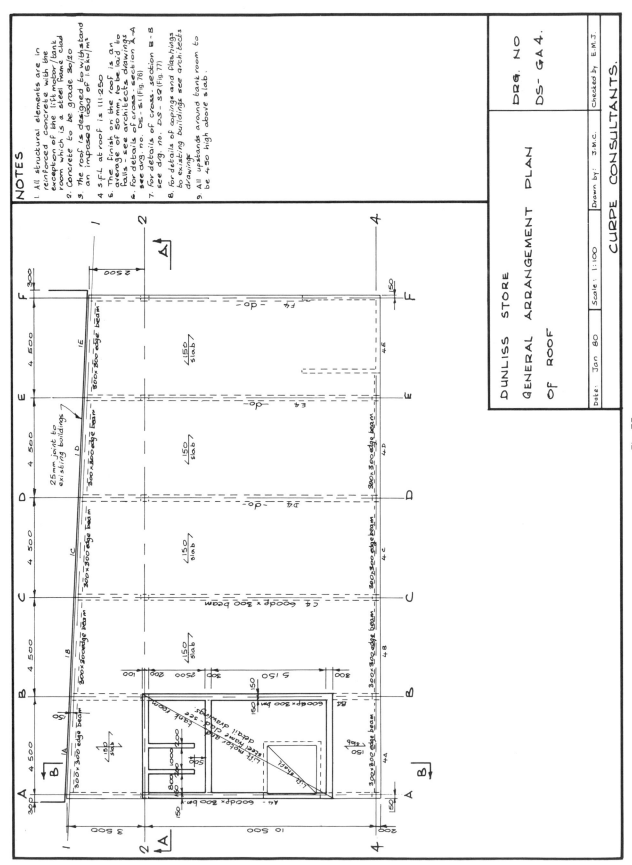

NOTES

1. All structural elements are in reinforced concrete with the exception of the lift motor/tank room which is a steel frame clad.

2. Concrete to be grade 30/20

3. The roof is designed to withstand an imposed load of 1.5 kN/m²

4. S.F.L at roof is 111.250

5. The finish on the roof is an average of 50mm, to be laid to falls - see architects drawings

6. For details of cross-section A-A see dwg. no. DS-S1 (Fig. 76)

7. For details of cross-section B-B see dwg. no. DS-S2 (Fig.77)

8. For details of copings and flashings to existing buildings see architects drawings

9. All upstands around tank room to be 450 high above slab.

DUNLISS STORE

GENERAL ARRANGEMENT PLAN

OF ROOF

| Date: Jan 80 | Scale: 1:100 | Drawn by: J.M.C. | Checked by E.M.J. |

DRG. NO
DS-GA4.

CURPE CONSULTANTS.

Fig. 75.

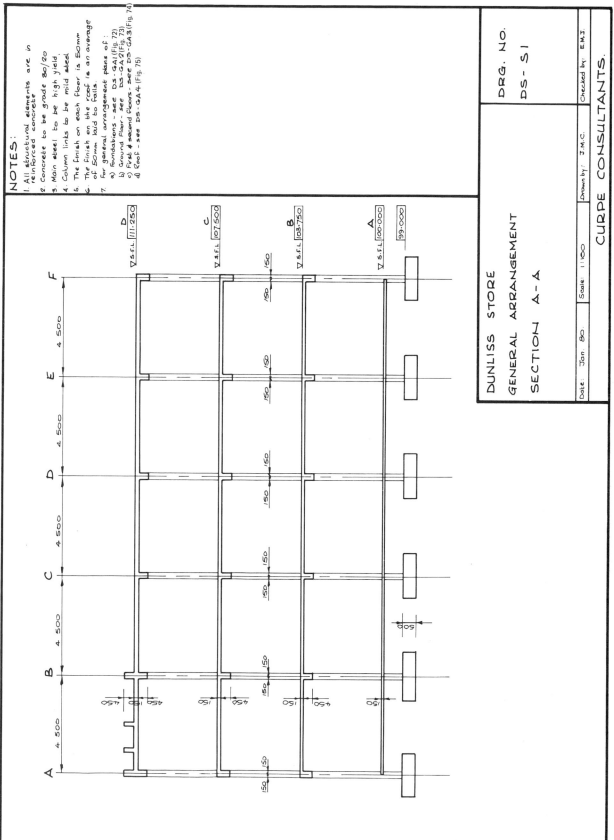

NOTES:
1. All structural elements are in reinforced concrete.
2. Concrete to be grade 30/20
3. Main steel to be high yield.
4. Column links to be mild steel.
5. The finish on each floor is 50mm.
6. The finish on the roof is an average of 50mm laid to falls.
7. For general arrangement plans of:
   a) Foundations - see DS-GA1(Fig.72)
   b) Ground Floor - see DS-GA2(Fig.73)
   c) First & second floors - see DS-GA3(Fig.74)
   d) Roof - see DS-GA4(Fig.75)

DRG. NO.
DS-S1

DUNLISS STORE
GENERAL ARRANGEMENT
SECTION A-A

Date: Jan. 80. Scale: 1:100 Drawn by: J.M.C. Checked by: E.M.I.

CURPE CONSULTANTS.

Fig. 76.

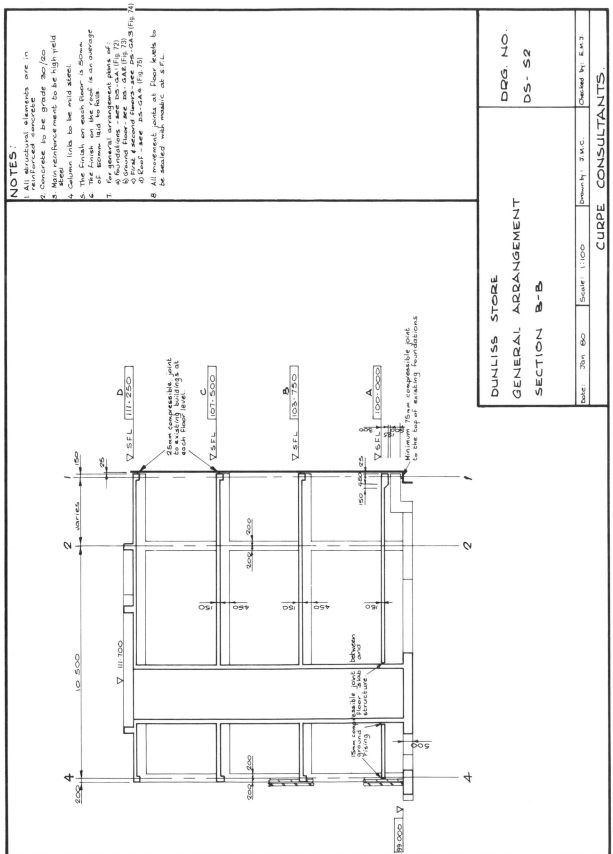

NOTES:

1. All structural elements are in reinforced concrete

2. Concrete to be grade 30/20

3. Main reinforcement to be high yield steel

4. Column links to be mild steel.

5. The finish on each floor is 50mm

6. The finish on the roof is an average of 50mm laid to falls.

7. For general arrangement plans of:
   a) Foundations - see DS-GA1 (Fig. 72)
   b) Ground floor - see DS-GA2 (Fig. 73)
   c) First & second floors - see DS-GA3 (Fig. 74)
   d) Roof - see DS-GA4 (Fig. 75)

8. All movement joints at floor levels to be sealed with mastic at s.f.l.

	DRG. NO.
DUNLISS STORE	DS-S2
GENERAL ARRANGEMENT	
SECTION B-B	Checked by: E.M.J.
Date: Jan 80	Scale: 1:100	Drawn by: J.M.C.	
	CURPE CONSULTANTS.

Fig. 77.

Slabs

INTRODUCTION

We have considered previously in Chapter Seven why reinforcement is required, where it should be positioned and the maximum and minimum areas of reinforcement as stipulated in CP110 *The Structural use of Concrete*. In this chapter we want to concentrate on how a reinforced concrete slab which is simply supported and spans in one direction only should be detailed. To do this task correctly the detailer must have certain information.

INFORMATION REQUIRED BY THE DETAILER

Outlines

The structural detailer must know the shape and size of the slab to be drawn. He can usually extract this information from the general arrangement drawings which would have been completed prior to detailing the various elements. For a slab it is usual to prepare a plan and a cross section with all dimensions and levels clearly shown.

Holes

Often holes are required in a slab, either for access or to permit the passage of electrical and/or mechanical services. It is important for these to be located correctly on the drawing so that the reinforcement can be positioned to suit. If the holes are small they may not affect the reinforcement but if large, special trimming reinforcement should be detailed to prevent cracks

developing at the corners of the openings.

The report referred to earlier, *Standard Reinforced Concrete Details*, gives recommendations on the treatment of holes in a solid slab. These refer to the positioning of holes, design implications and reinforcement details. The detailer should appreciate at least the first and last of these, which are reproduced below.

Positioning of holes
(*a*) No hole should be closer than its width to an unsupported edge (*see* Fig. 78).

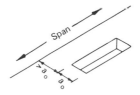

Fig. 78. *No hole should be closer than its width to an unsupported edge.*

(*b*) The maximum width of hole measured transversely to the main reinforcement should not be greater than 1,000 mm.

(*c*) In any 4 m width of slab, measured at right angles to the main reinforcement, not more than

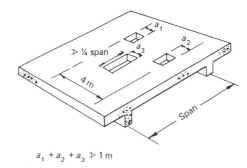

$a_1 + a_2 + a_3 > 1$ m

Fig. 79. *In any 4 m width of slab, no more than a quarter should be removed.*

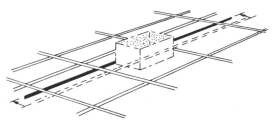 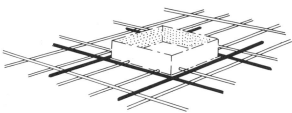

Fig. 80. *A hole with a width at right angles to the span not greater than 200 mm, showing the reinforcements moved to one side.*

Fig. 81. *A hole with a width not greater than 500 mm, where the reinforcement has been cut and replacement reinforcement provided.*

one-quarter of the slab should be removed (*see* Fig. 79). For slabs more than 4 m wide, the permissible width of holes should be reduced proportionately.

(*d*) The length of hole measured parallel to the main reinforcement should not be greater than one-quarter of the span.

Reinforcement details

A hole which has no structural significance, i.e. where the designer considers that it will not affect the structural behaviour of the slab, may be trimmed by one of the following methods, depending upon the size of the opening:

(*a*) holes with a width at right angles to the span not greater than 200 mm, where the reinforcement in the slab is not cut (*see* Fig. 80);

(*b*) holes with a width not greater than 500 mm, where the bars are cut and replacement reinforcement provided (*see* Fig. 81);

(*c*) holes with a width not greater than 1,000 mm, where replacement reinforcement is provided and additional bars are placed in the top of the slab (*see* Fig. 82); may be trimmed as indicated below.

In all cases the trimming reinforcement should be placed as close to the edge of the hole as practical.

All diagrams relate to the main reinforcement in the bottom of slabs. Top reinforcement should be treated in the same manner as the bottom steel.

(*a*) *Width of hole not greater than 200 mm*. The reinforcement which interferes with the hole may be moved to one side provided that:

(*i*) the slab thickness is not more than 250 mm;

(*ii*) the resultant spacing of bars does not exceed three times the effective depth of the slab.

If either of these conditions is not satisfied, the trimming reinforcement should be as in (*b*).

(*b*) *Width of hole not greater than 500 mm*. The main reinforcement which interferes with the hole should be cut and replaced by bars of the same size placed evenly on all sides of the hole. Additional reinforcing bars will be required if an uneven number of bars are cut.

If the distribution reinforcement interferes with the hole, it should be cut.

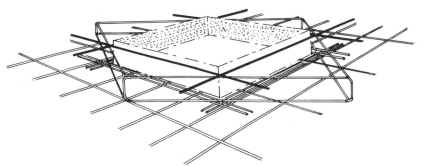

Fig. 82. *A hole with a width not greater than 1,000 mm, where the reinforcement has been cut and replacement reinforcement provided.*

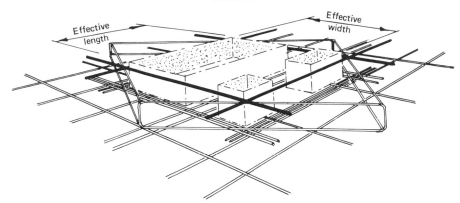

Fig. 83. *When the distance between holes is less than one and half times the width of the largest opening, the group should be treated as one hole.*

All replacement bars should extend an anchorage length beyond the edges of the hole.

(*c*) *Width of hole not greater than 1,000 mm.* The main reinforcement should be treated in the same manner as in (*b*), but with additional replacement bars of the same size placed on each side of the hole in the top of the slab.

Diagonal reinforcement in the form of links or straight bars supported by chairs or links of the same size as the replacement reinforcement should be placed at the corners of the hole. The effective length of the diagonal bars should be at least two anchorage lengths.

Groups of holes
When the distance between holes is less than one and a half times the width of the largest opening, the group should be treated as a single hole with an effective width and length as indicated in Fig. 83. The trimming reinforcement for groups of holes should be as in (*b*) or (*c*) above, depending upon the effective width. Wherever possible, bars should be moved to the sides of holes. In all cases one additional replacement bar of the same size as the main reinforcement should be placed on each side of the group of holes.

Direction of span
In a slab which is supported on two sides only it is obvious that it will span between those supports. Where all edges are supported it is still possible to design the slab as simply supported spanning in one direction if the length/breadth ratio exceeds 2. In this case the slab would span across the shorter dimension. The detailer needs to know the direction of span because this determines the direction of the reinforcement, i.e. the main reinforcement is positioned parallel to the direction of the span and the secondary reinforcement normal to the direction of the span.

Indicating the reinforcement
The reinforcement, which will consist of many bars, is indicated by drawing one bar of the group in full, with an extent line drawn through this bar extending to the short lines which show the position of the first and last bars.

These positions, together with the bar drawn, are indicated using arrowheads. The extent line is normally produced so as to be outside the member, where it can conveniently be labelled in accordance with the standard format outlined in Chapter Two.

Areas of reinforcement
The main reinforcement is designed by the structural engineer and the experienced detailer should be able to extract the reinforcement required from the engineer's calculations. Where the minimum area is necessary the detailer could calculate this as shown in Chapter Seven, which also illustrated how the amount of secondary reinforcement would be determined.

Rules for detailing
CP110, under Section 3.11.7, gives recommendations for the curtailment and anchorage of reinforcement. In Subsection 3.11.7.3, *Simplified*

Rules for Curtailment of Bars in Slabs, the following rule is given for simply supported slabs.

"At least 50 per cent of the tension reinforcement provided at mid-span should extend to the supports and have an effective anchorage of twelve times the bar diameter past the centre of the support. The remaining 50 per cent should extend to within 0.08 *l* [the span of the slab] of the support."

Anchorage bond lengths

For laps CP110, Clause 3.11.6.5, states:

"When bars are lapped, the length of the lap should at least equal the anchorage length [given in Table 22] required to develop the stress in *the smaller* of the two bars lapped, except that for deformed bars in tension the length of the lap should be 25 per cent *greater* than the anchorage length required for the *smaller bar*. The length of the lap provided, however, should neither be less than twenty-five times the bar size plus 150 mm in tension reinforcement."

Table 22 gives the anchorage bond lengths for mild steel and type 1 high yield deformed bars. An explanation of the types of reinforcement may be found in Appendix E of CP110.

Miscellaneous

Other information which should appear on the detailed drawing is:

(*a*) the imposed loading that the slab has been designed to withstand, obtained from the designer or his calculations;

(*b*) the grade of concrete and the maximum size of aggregate to be used in the mix, again obtained from the designer;

(*c*) the cover to the reinforcement, obtained from Table 19 of CP110 or the designer's calculations.

The method of detailing a slab and scheduling the reinforcement is best understood by an illustration (*see* Figs. 84, 85 and 86).

TABLE 22. ANCHORAGE BOND LENGTHS

Plain round mild steel bars ($f_y = 250$ N/mm²):

Grade of concrete	8 mm		10 mm		12 mm		16 mm		20 mm		25 mm		32 mm	
	C	T	C	T	C	T	C	T	C	T	C	T	C	T
20	310*	365	350*	455	395	545	525	725	655	910	820	1,135	1,050	1,450
25	310*	350*	350*	400*	390*	470	470*	625	580	780	725	975	925	1,245
30	310*	350*	350*	400*	390*	450*	470*	580	550*	725	650*	910	830	1,160

Deformed high yield bars ($f_y = 410$ N/mm²):

Grade of concrete	8 mm		10 mm		12 mm		16 mm		20 mm		25 mm		32 mm	
	C	T	C	T	C	T	C	T	C	T	C	T	C	T
20	310*	420	365	525	435	630	580	840	725	1,050	905	1,315	1,155	1,680
25	310*	380	350*	470	390*	565	505	755	635	940	790	1,115	1,010	1,505
30	310*	350*	350*	410	390*	490	470*	650	565	815	705	1,015	900	1,300

NOTES:

(*a*) Dimensions marked * are the minimum allowed under the rules (20 × diameter) + 150 mm for compression and (25 × diameter) + 150 for tension.

(*b*) The other values assume that the stress in the bar at the point from which the length is being measured is a maximum.

(*c*) When two bars of different diameters are being lapped the lap length given should correspond to the smaller bar.

(*d*) C = compression: T = tension.

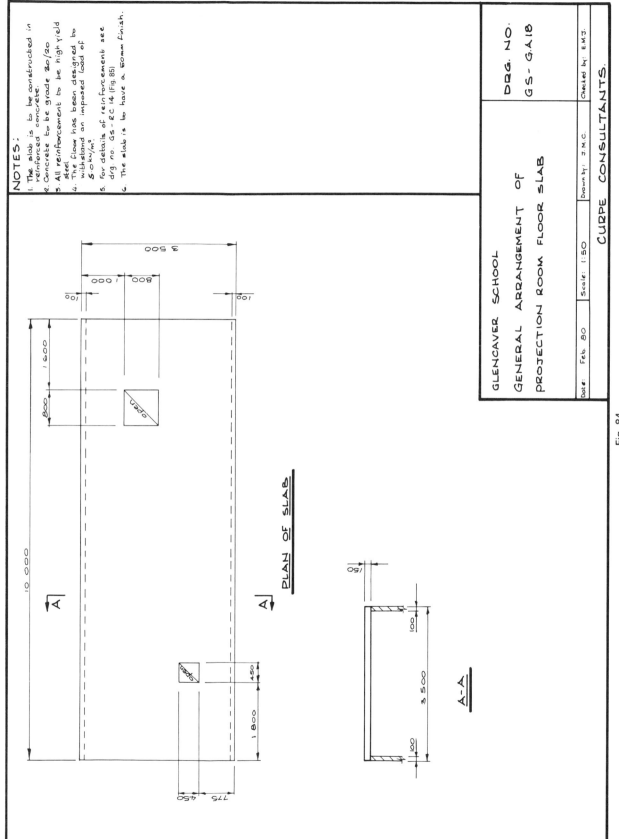

NOTES:

1. The slab is to be constructed in reinforced concrete.
2. Concrete to be grade 20/20
3. All reinforcement to be high yield steel
4. The floor has been designed to withstand an imposed load of 5.0 kN/m²
5. For details of reinforcement see drg. no. GS-RC 14 (Fig.85)
6. The slab is to have a 50mm finish.

PLAN OF SLAB

A-A

GLENCAVER SCHOOL			DRG. NO.
GENERAL ARRANGEMENT OF			GS- G.A.18
PROJECTION ROOM FLOOR SLAB			
Date: Feb. 80	Scale: 1:50	Drawn by: J.M.C.	Checked by: E.M.J.

CURPE CONSULTANTS.

Fig. 84.

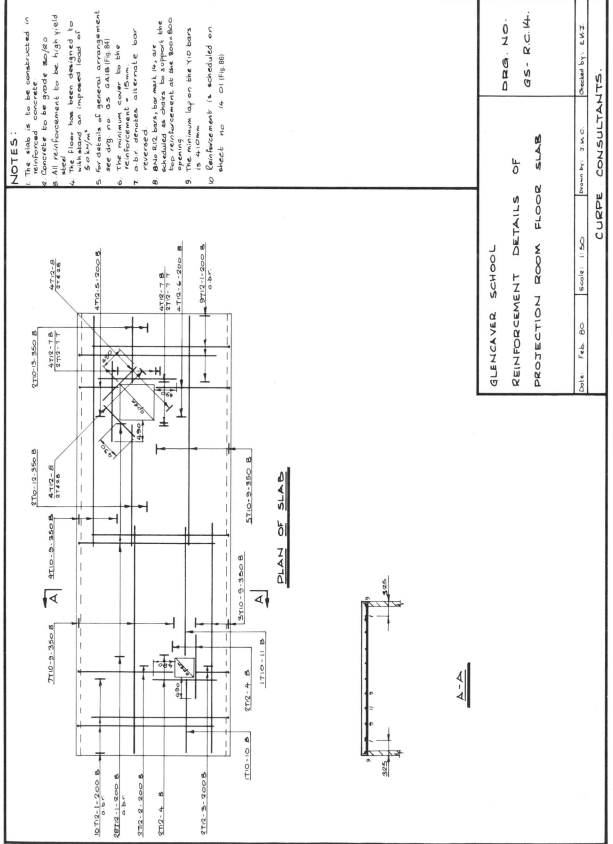

NOTES:
1. The slab is to be constructed in reinforced concrete.
2. Concrete to be grade 30/20
3. All reinforcement to be high yield steel
4. The floor has been designed to withstand an imposed load of 5·0 kN/m²
5. For details of general arrangement see drg. no. GS GA18 (Fig.84)
6. The minimum cover to the reinforcement = 15mm.
7. a.b.r. denotes alternate bar reversed
8. 8No R12 bars, bar mark 14, are scheduled as chairs to support the top reinforcement at the 800×800 opening.
9. The minimum lap on the Y10 bars is 410mm.
10. Reinforcement is scheduled on sheet no 14·01 (Fig.86)

PLAN OF SLAB

A-A

GLENCAVER SCHOOL
REINFORCEMENT DETAILS OF
PROJECTION ROOM FLOOR SLAB

Date: Feb. 80. | Scale: 1:50 | Drawn by: J.W.C. | Checked by: E.W.J.

DRG. NO.
GS-R.C.14.

CURPE CONSULTANTS.

Fig. 85.

ILLUSTRATION

We shall assume the following.

(*a*) A simply supported slab has been designed and the amount of main reinforcement required has been calculated to be 12 mm diameter bars of high yield reinforcement at 200 mm centres and the secondary, or distribution, steel to be 10 mm bars of high yield reinforcement at 350 mm centres.

(*b*) The concrete is to be grade 30 with a maximum aggregate size of 20 mm.

(*c*) The minimum cover to the reinforcement is to be 15 mm.

(*d*) The imposed loading the slab has been designed to withstand is 5.0 kN/m².

(*e*) The drawing for the outline has to be separate from the drawing for the reinforcement details.

Figures 84 and 85 and the schedule in Fig. 86 illustrate the work involved for the structural detailer.

CURPE CONSULTANTS

Bar Schedule ref. `0 1 4`   `0 1`   Rev. ☐

Site Ref. Glencaver School   Date:

Member	Bar mark	Type & size	No. of mbrs.	No. in each	Total	Length of each bar mm	Shape code	A mm	B mm	C mm	D mm	E/r mm
Slab	1	T12	1	47	47	3275	34	3160				
	2	T12	1	2	2	2325	34	2215				
	3	T12	1	2	2	825	34	730				
	4	T12	1	4	4	1450	20	STRAIGHT				
	5	T12	1	4	4	1050	34	955				
	6	T12	1	4	4	1750	34	1650				
	7	T12	1	12	12	1800	20	STRAIGHT				
	8	T12	1	8	8	1000	20	STRAIGHT				
	9	T10	1	19	19	5175	20	STRAIGHT				
	10	T10	1	1	1	1725	20	STRAIGHT				
	11	T10	1	1	1	2925	20	STRAIGHT				
	12	T10	1	2	2	2775	20	STRAIGHT				
	13	T10	1	2	2	1525	20	STRAIGHT				
	14	R12	1	8	8	825	83	250	95	250		

Fig. 86. *Bar schedule for Figs. 84 and 85.*

SELF-ASSESSMENT QUESTION

A simply supported slab is 175 mm thick and spans 3.200 m between the centres of supporting block walls which are 150 mm thick. The slab is 5.750 m long with a centrally positioned opening measuring 900 mm square. The slab is to be inside a building and the engineer has decided to use a 25 grade concrete.

The reinforcement has been designed and found to be 12 mm diameter bars of high yield reinforcement at 150 mm centres for the main reinforcement; the secondary reinforcement is minimal.

Produce a drawing and a schedule showing all outline and reinforcement details.

Columns

DEFINITION

CP110 *The Structural Use of Concrete* does not define a column, but in Clause 3.8.1.1. it defines a wall as follows: "a vertical load bearing concrete member whose greatest lateral dimension is more than four times its least lateral dimension ...". By implication, therefore, we may define a column as a vertical load bearing concrete member whose greatest lateral dimension is not more than four times its least lateral dimension.

It is the purpose of this chapter to outline how a column should be detailed and to illustrate this with an example.

DETAILING

Outlines
The Report recommends that columns should be detailed in elevation. This means that the detailer must prepare clear outline drawings of the columns, showing sufficient sections to indicate clearly the cross-sectional dimensions. All floor levels and beam intersections should be drawn since these have a major bearing on the bar lengths and shapes. These outlines should be drawn to a scale of 1:20 and may be extracted from the general arrangement drawings. The column sections should incorporate the grid lines to allow for simple orientation of the column.

Reinforcement
Two very different forms of reinforcement are employed in a column, the longitudinal main reinforcement and the transverse linkage reinforcement. The rules relating to the maximum and minimum areas of this reinforcement have been outlined previously in Chapter Seven. The actual amount of reinforcement to be used will normally lie between these values and should be obtained from the design engineer's calculations.

Main reinforcement
When detailing the main reinforcement in a column the detailer is faced with two problems:

    (*a*) how to show the bars;
    (*b*) how to splice the bars.

How to show the bars. The main bars should be shown in elevation by drawing one bar in full with the remainder indicated by short lines with a dimension line across the set. If there is more than one type of main bar each type should be drawn and indicated separately.

How to splice the bars. There are at least five methods which have been used to give continuity to the main reinforcement in a column. The bars may be:

    (*a*) welded;
    (*b*) threaded;
    (*c*) mechanically spliced;
    (*d*) lapped;
    (*e*) spliced using dowel bars.

Alternatives (*a*) and (*b*) are usually limited to shop prepared precast work where the fine tolerances and controlled conditions necessary can be achieved.

Option (*c*) is not normally suitable where tension or torsion has to be resisted. However, compressive forces may be transmitted by end bearing of square cut bars, held in concentric contact by a specially designed and manufactured steel sleeve. Spacers inside the sleeve can be used where two bars of differing diameters are to be joined. This type of joint can be more expensive than a straightforward lap, but where the reinforcement is congested and lapping difficult it can prove very useful.

Easily the most common method of achieving continuity is (d), lapping. The position of the lap should be immediately above the floor/beam/column junction, enabling a storey height of the column to be poured with the column bars projecting a lap length above the kicker. The bars lapping with these starter bars should be cranked to form a lap (see Fig. 87). This also provides the most satisfactory detail from a design con-

sideration since the reinforcement is in the best position to withstand any bending stresses at the ends of the column. CP110 in Clause 3.11.4.3 requires that all bars or groups of bars within a compression zone should be within 150 mm of a restrained bar (see Fig. 88).

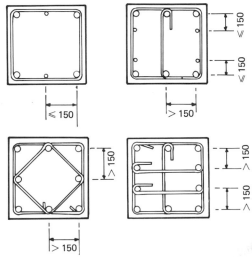

Fig. 88. *All bars or groups of bars within a compression zone should be within 150 mm of a retained bar.*

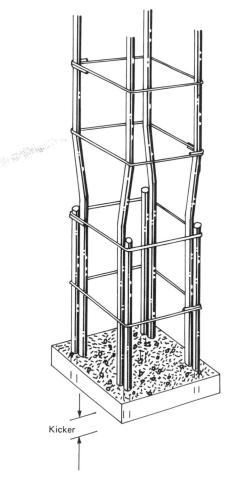

Fig. 87. Bars lapping with starter bars.

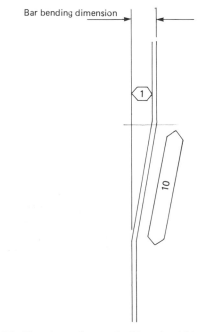

Fig. 89. *The slope of a cranked bar should be 1 : 10.*

Where the reinforcement consists of ribbed or square twisted bars it is important to remember that the overall diameter of the bar is about 10 per cent greater than its nominal diameter. This must be taken into account when scheduling the offset of the cranked bar and a minimum tolerance of 10 mm should be provided between the reinforcement cages. The slope of the inclined portion of a cranked bar should be 1:10, as shown in Fig. 89.

When two bars of differing diameter are being lapped, the anchorage bond length (*see* Table 22) of the smaller bar should be given.

The fifth option, (*e*)—splicing by dowel bars—may be used in preference to lapping but this leads to an uneconomical use of reinforcement since two lap lengths are required rather than one. Where the column changes section, or a face or faces are offset, separate dowel bars should be provided. This avoids cranking the column bars at both ends. A typical detail is shown in Fig. 90, where the special links which are wired to the lower reinforcement cage and are used to locate the dowel bars should be noted. These dowel bars should be cast at least an anchorage bond length into the lower column.

Links

These are the bars which form the transverse reinforcement and are sometimes called binders or stirrups. In Chapter Seven we dealt with their size and spacing but not their arrangement.

Link arrangement. Clause 3.11.4.3 of CP110 states that "links should be so arranged that every corner and alternate bar or group in an outer layer of reinforcement is supported by a link passing round the bar and having an included angle of not more than 135°". Not all the links have to be of the closed variety, provided they comply with Clause 3.11.6.4.

"A link may be considered to be fully anchored if it passes round another bar of at least its own size through an angle of 90° and continues beyond for a minimum length of eight times its own size, or through 180° and continues for a minimum length of four times its own size" (*see* Fig. 91).

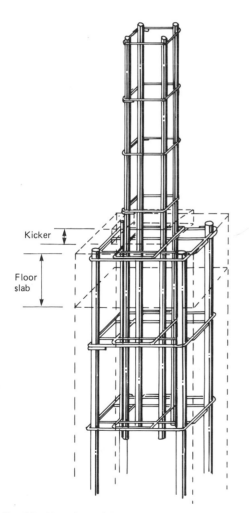

Fig. 90. *Use of special links to locate dowel bars.*

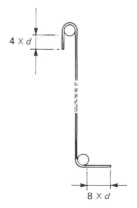

Fig. 91. *A fully anchored link.*

Care must be taken to ensure the correct positioning of links at the upper and lower offset points in a cranked bar. This is to resist the horizontal or bursting forces (*see* Fig. 92).

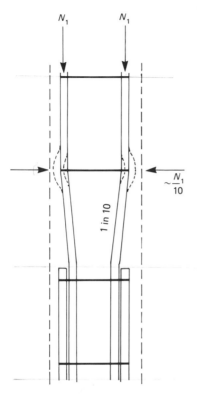

Fig. 92. *Correct positioning of links at upper and lower offset points in a cranked bar.*

Link detailing. Each set of links should be indicated by one link, or group, drawn in full, the first and last being shown by a short line and a dimension or extent line drawn through the link or group to these short lines. The fixing positions of the links should be clearly drawn on the sections through the column.

Link scheduling. Since a longitudinal bar will not usually fit tightly into the corner of a link, where it is enclosed by two or more links the internal width of the link should allow a tolerance to enable the longitudinal bar to be fixed in its correct position.

The internal bending dimension of a link should be calculated as follows:

Column cross-sectional dimension $- 2 \times$ (nominal cover + Tolerance + Link diameter + 5 mm)

The 5 mm figure comes from the Table of Tolerances (*see* Table 5, page 20).

If the cover is greater than 40 mm to the main steel and a normal structural concrete is being used, supplementary reinforcement consisting of either a wire fabric not lighter than 0.5 kg/m² (2 mm diameter wires at not more than 100 mm centres) or a continuous arrangement of links at not more than 200 mm centres should be incorporated in the concrete cover at a distance of not more than 20 mm from the face (*see* CP110, Clause 10.5). This is to help prevent the concrete spalling when exposed to fire.

ILLUSTRATION

The illustration shown in Fig. 93 is typical of the work required from a structural detailer. A number of points should be made.

(*a*) Since the outline is simple, both it and the reinforcement have been shown on the one drawing.

(*b*) Bar mark 9 is required since the bars in the centre of the column on grid line 6 are more than 150 mm from the corner bars. The converse is true of the bars on grid line B.

(*c*) Note the arrangement of splice bars where the column changes section at level 110.500.

(*d*) Where bars are left projecting above a floor level and lapping with bars of differing diameters the longest lap has been kept as a constant so that only one type of bar has to be drawn and scheduled.

The bar schedule is shown in Fig. 94.

SELF-ASSESSMENT QUESTIONS

1. Draw two alternative link arrangements for each of the columns shown in Fig. 95.

2. The top level of a column foundation is 2.500 and there are six 32 mm diameter bars of high yield reinforcement left protruding a lap length above a 75 mm kicker. The column is 450 mm square from level 2.50 to level 7.50 where it changes to 300 mm square central on the lower column. This section passes through a floor intersection at 11.25 and the roof slab is at level 15.00.

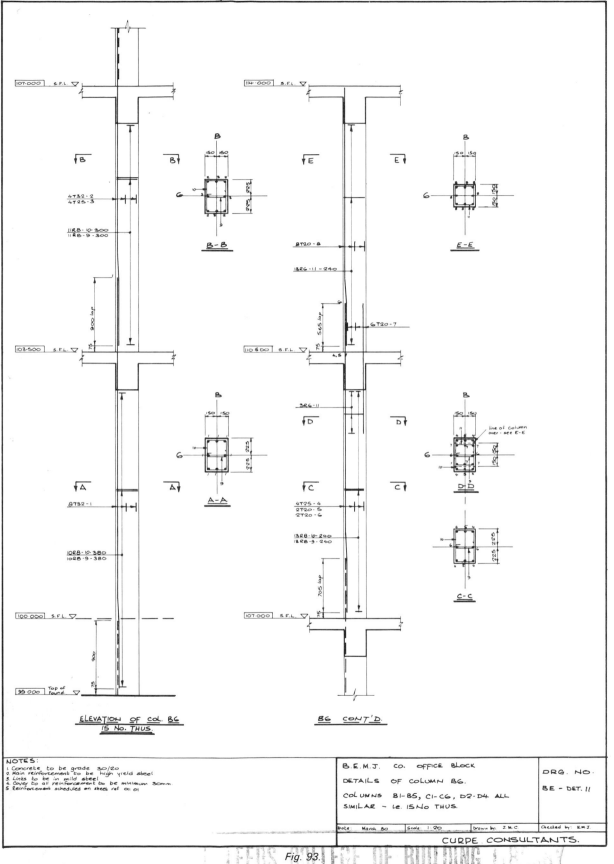

ELEVATION OF COL. B6
15 No. THUS.

B6 CONT'D.

NOTES:
1. Concrete to be grade 30/20
2. Main reinforcement to be high yield steel
3. Links to be in mild steel
4. Cover to all reinforcement to be minimum 30mm.
5. Reinforcement scheduled on sheet ref. 01.01

B.E.M.J. CO. OFFICE BLOCK

DETAILS OF COLUMN B6.

COLUMNS B1-B5, C1-C6, D2-D4 ALL
SIMILAR - i.e. 15No. THUS.

DRG. NO.

BE - DET. 11

| Date: March 80 | Scale: 1:20 | Drawn by: J.M.C. | Checked by: E.M.J. |

CURDE CONSULTANTS.

Fig. 93.

CURPE CONSULTANTS

Bar Schedule ref. ☐ 0 / 1 / ☐ 0 / 1 ☐ Rev. ☐

Site Ref. B.E.M.J Co. Office Block   Date:

Member	Bar mark	Type & size	No. of mbrs.	No. in each	Total	Length of each bar mm	Shape code	A mm	B mm	C mm	D mm	E/r mm
COLUMN B6	1	T32	15	8	120	5400	41	900	350		70	
Cols B1-B5, C1-C6,	2	T32	15	4	60	4200	41	900	350		70	
D2-D4 Similar	3	T25	15	4	60	4200	41	900	300		65	
	4	T25	15	4	60	3350	41	705	300		65	
	5	T20	15	2	30	3350	41	705	300		50	
	6	T20	15	2	30	4050	41	705	300		50	
	7	T20	15	6	90	1725	20	STRAIGHT				
	8	T20	15	8	120	3350	41	565	300		45	
	9	R8	15	34	510	475	85	100	225	75		
	10	R8	15	34	510	1300	60	360	210			
	11	R6	15	16	240	1000	60	215	215			

Fig. 94. Bar schedule for Fig. 93.

The reinforcement in the second lift is 6Y25 mm bars and 6Y20 mm bars in the top lift with nominal mild steel links in each case. If the floors are 150 mm thick with beams 300 mm wide × 500 mm deep and the minimum cover to the reinforcement is 30 mm, draw and detail the column together with the reinforcement schedule.

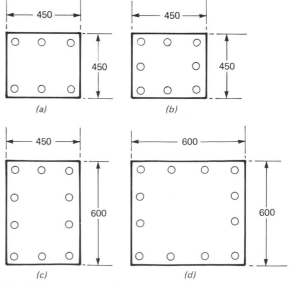

Fig. 95. Cover to main reinforcement is 40 mm in each case.

Beams

INTRODUCTION

We have now considered all the basic reinforced concrete elements with the exception of simple beams. The more complex elements, including continuous beams, will be dealt with in *Structural Detailing III*.

The detailing of beams causes particular problems due to the larger diameter bars employed and the confined space into which they must be fixed. Careful co-ordination between the designer and the detailer can eliminate virtually all these problems, enabling the bending and fixing of the reinforcement to be carried out without the delays sometimes encountered due to careless detailing.

It is the purpose of this chapter to outline how simple beams should be detailed and to illustrate this work with an example.

DETAILING

Outlines

Beams are usually detailed in elevation, with sufficient sections to show the cross-sectional dimensions on one drawing and the reinforcement details and locations on another. Where the beams are wide and flat it may be easier to show these details on a plan rather than an elevation.

The outline must show the relationship with all other structural members such as intersecting columns, beams and slabs, as well as the position and size of any service holes. Using the established grid and level system, the location should be clearly indicated.

All this information may be obtained from the general arrangement drawings.

Reinforcement

With beams, as with columns, it is obvious that two different forms of reinforcement are employed, i.e. the main longitudinal bars and the transverse links. The rules relating to the maximum and minimum areas of this reinforcement have been outlined already in Chapter Seven.

Notwithstanding these rules, the amount of reinforcement to be used in any particular beam should be determined from the engineer's calculations.

Main reinforcement

Layers of bars. The main or longitudinal reinforcement in a beam may consist of one or more layers. Each layer should be indicated by a single line in elevation, with the ends of each bar shown by a tick which is labelled with the bar mark. These layers are often held apart by utilising spacer bars to enable the concrete to flow between the layers. These spacer bars should be detailed and scheduled, and are normally cut from mild steel without regard for the high yield main bars.

Lapping of bars. At a lap position the larger bars normally used in a beam are generally cranked (*see* Fig. 87). The other methods outlined in Chapter Ten may also be used, but methods (*d*) and (*e*), i.e. lapping by cranking or splicing using dowel bars, are the most common, with the former being easily the most widely used. Where the bars are of 12 mm diameter or less there is no necessity to crank them at lap positions.

The positioning of laps in a beam is most important and, where possible, should be in an area of compression rather than tension. These positions should be chosen with the guidance of the structural engineer.

Special consideration must be given to any positions where layers of bars are to be lapped. This should be done in such a manner as to avoid congestion or clashing of the reinforcement. If this is a recurring problem throughout the structure a regular system should be established and adhered to by all detailers working on the project.

Lengths for anchorage. For laps CP110, Clause 3.11.6.5, states that:

"When bars are lapped, the length of the lap should at least equal the anchorage length [given in Table 22] required to develop the stress in *the smaller* of the two bars lapped, except that for deformed bars in tension the length of the lap should be 25 per cent *greater* than the anchorage length required for the *smaller* bar. The length of the lap provided, however, should neither be less than twenty-five times the bar size plus 150 mm in tension reinforcement nor be less than twenty times the bar size plus 150 mm in compression reinforcement."

Links

These are the bars which are placed around the main longitudinal reinforcement and at right angles to it to form the transverse reinforcement. Their function in a beam is usually to resist shear or torsional stresses. Sometimes they are referred to as binders or stirrups.

The links may be indicated by one of the following methods:

(*a*) draw the first and last bars or groups in full and draw the extent line outside the concrete outline with the bars labelled on this line;

(b) draw one link or group in full, with a dimension or extent line drawn through the link or group extending to the short lines which show the position of the first and last bars.

The former method avoids congestion within the outline on the reinforcement drawing and is therefore preferable.

Shape of links. These should be chosen from the standard shapes given in BS4466 (*see* Fig. 13). Shape codes 60, 73 and 81 are commonly used, with shape code 73 being particularly useful. This allows convenient fixing of the main reinforcement since the top is open, but separate closing bars may be required which have to be detailed and scheduled. Where the beam is narrow there may be insufficient space between the bar ends to permit shape 73 to be used.

It is important for the detailer to realise that bars on site cannot be positioned just as neatly as can be indicated on the drawing. When a main longitudinal bar is embraced by two or more links the internal width of the links must allow sufficient tolerance to enable the main bar to be located. This is due to the fact that the main bars do not fit tightly into the corners of the link (*see* Fig. 96).

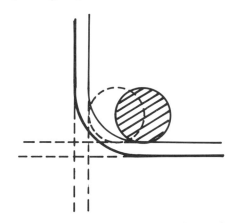

Fig. 96. *Bars on site cannot be positioned as neatly as can be indicated in a drawing.*

Size of links. Mild steel bars of 6 mm, 8 mm, 10 mm or 12 mm diameter are commonly specified for links. This is because larger diameter bars are difficult to bend and mild steel is preferred to high yield since the internal radius for bending for the former is two diameters while for the latter it is three diameters. This means that the problem illustrated in Fig. 96 is alleviated with mild steel.

In beams more than 1 m deep, links of less than 12 mm diameter are particularly susceptible to buckling. The remedy is either to use larger diameter links, with the problem of bending, or to provide lacer bars on the face of the beam

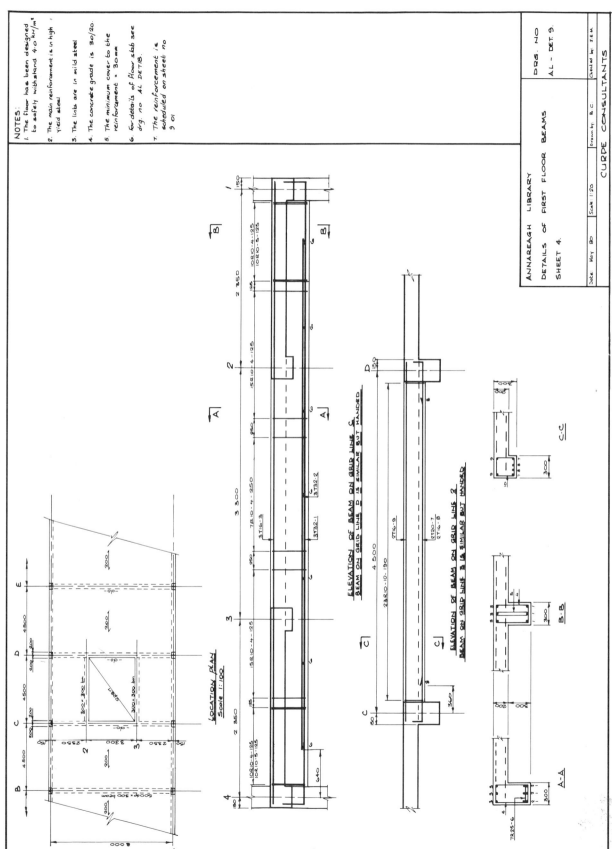

Fig. 101.

CURPE CONSULTANTS	Project ANNAREAGH LIBRARY			Job ref CC 107	
46, ORBURN ROAD DUNFIELD	Part of structure 1st FLOOR SLAB			Calc sheet no rev / /	
	Drawing ref	Calc by BC	Date MAY 80	Check by	Date
Ref	Calculations			Output	

SUMMARY OF BEAMS ON GRID LINES C&D

Top Reinf - 3T16 ∅ bars

Bottom Reinf - 6T32 ∅ bars

Links - R10 in pairs at 125mm % for 1·20m from suppt.

   R10 singly at 125mm % for next 1·80m

   R10 " " 250 mm % for central 2m.

SUMMARY OF BEAMS ON GRIDLINES 2&3.

Top Reinf - 2T16 ∅ bars

Bottom Reinf - 2T20 ∅ bars + 2T16 ∅ bars

Links - R10 singly at 190mm % throughout

Fig. 102. *Structural engineer's calculations for Fig. 101.*

CURPE CONSULTANTS

Bar Schedule ref. | 0 | 0 | 9 | | 0 | 1 | Rev. ☐

Site Ref.   *Annareagh Library*   Date:

Member	Bar mark	Type & size	No. of mbrs.	No. in each	Total	Length of each bar mm	Shape code	A mm	B mm	C mm	D mm	E/r mm
Beam on Grid Line C, Beam on Grid Line b Similar	1	T32	2	3	6	8575	38	275	8210			
	2	T32	2	3	6	6700	20	STRAIGHT				
	3	T16	2	3	6	8175	20	STRAIGHT				
	4	R10	2	57	114	1425	60	405	205			
	5	R10	2	20	40	1250	81	405	75			
	6	R25	2	7	14	200	20	STRAIGHT				
Beam on Grid line 2, Beam on Grid Line 3 Similar	7	T20	2	2	4	5000	38	150	4710			
	8	T16	2	2	4	3775	20	STRAIGHT				
	9	T16	2	2	4	4700	20	STRAIGHT				
	10	R10	2	23	46	1025	60	205	205			

Fig. 103. *Bar schedule for Fig. 101.*

SELF-ASSESSMENT QUESTION

Assume a layout similar to that shown in Fig. 101, except that the grid lines A–B, B–C, C–D, etc. are 5.0 m apart, 1–4 are 8.50 m apart and 2–3 are 3.50 m apart, centrally positioned.

Figure 104 shows the summary of the structural engineer's calculations.

Detail the beams on grid lines C, D, 2 and 3, together with a reinforcement schedule, assuming beams on grid lines C and D are 575 mm deep and beams on grid lines 2 and 3 are 350 mm deep.

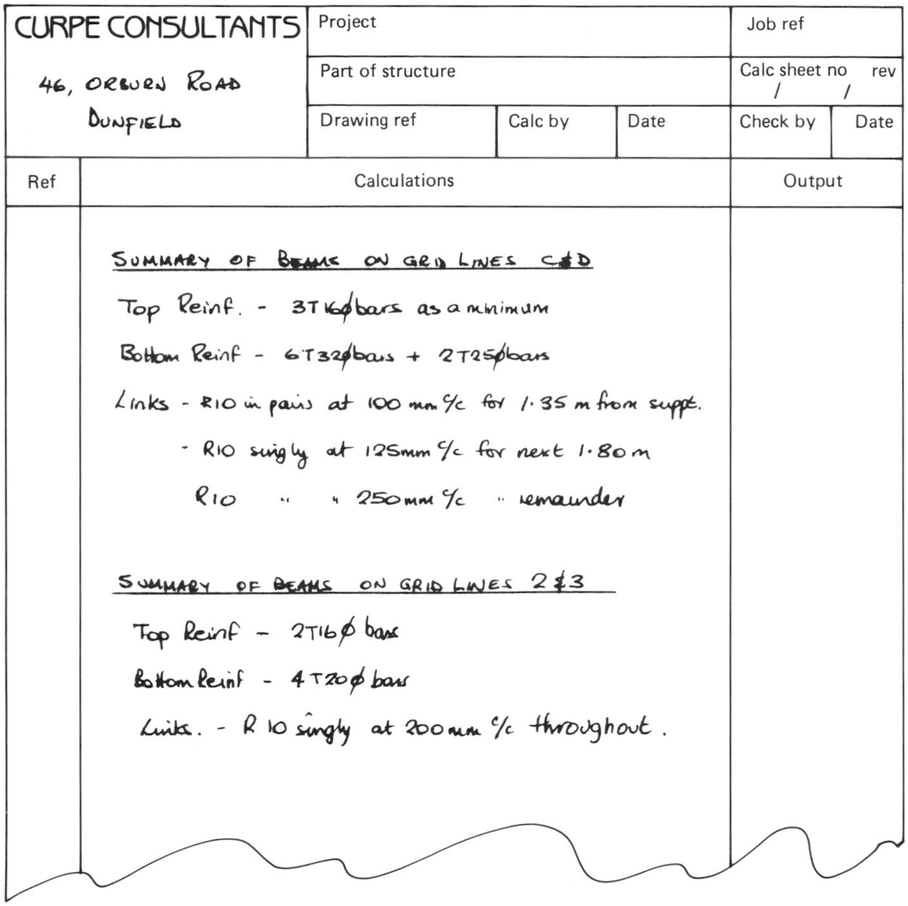

CURPE CONSULTANTS 46, ORBURN ROAD DUNFIELD	Project			Job ref	
	Part of structure			Calc sheet no /	rev /
	Drawing ref	Calc by	Date	Check by	Date
Ref	Calculations			Output	

SUMMARY OF BEAMS ON GRID LINES C&D

Top Reinf. - 3T16∅ bars as a minimum

Bottom Reinf - 6T32∅ bars + 2T25∅ bars

Links - R10 in pairs at 100 mm ℅ for 1.35 m from supp.

　　　 - R10 singly at 125mm ℅ for next 1.80 m

　　　　R10　 " 　" 250mm ℅ " remainder

SUMMARY OF BEAMS ON GRID LINES 2&3

Top Reinf - 2T16∅ bars

Bottom Reinf - 4T20∅ bars

Links. - R10 singly at 200 mm ℅ throughout.

Fig. 104.

Glossary of Terms

Angle cleat. A small bracket of steel angle section used for steel beam and stanchion connections.

Beam. A horizontal load bearing member whose vertical dimension is usually large in comparison with the horizontal dimension.

Bearing pressure. The load on a bearing surface divided by the area of that surface.

Bevelled or tapered washer. A steel wedge shaped washer used at the tapered flange of a rolled steel joist or channel.

Black bolts. Bolts covered with scale, i.e. black iron oxide.

Blinding. A layer of lean concrete put down on soil to seal it and provide a clean working surface.

Boom or chord. The horizontal members, top and bottom, of a built-up lattice girder.

Borehole. A hollow cylinder driven into the ground in order to extract soil samples which reveal information about the soil strata.

Bracing. Members, usually diagonal, used to stiffen a structure, often against wind forces.

Bush-hammering. The removal of the outer skin of a concrete surface using a light percussive tool.

Casing. The concrete surround often given to steel beams and stanchions.

Chairs. Reinforcement bars bent in such a way as to support the reinforcement in the top of a concrete slab by resting on the bottom reinforcement.

Checker. A leader of a section in a drawing office who checks the structural drawings.

Clerk of works. Often referred to as the C.O.W., he is the official appointed on the behalf of the client to ensure that the contractor's building work complies with that specified.

Column. A vertical load bearing member of relatively small cross section, made from reinforced concrete.

Consultant or consulting engineer. The specialist who acts on behalf of the client or the architect to produce the complete design of the building.

Contractor. The builder who signs a contract to produce the required structure to a given specification for an agreed payment within a specified time.

Countersunk. A conical enlargement at the end of a hole to receive the head of a bolt or screw so that there is no projection above the surface.

Cover. The thickness of concrete between the concrete face and the nearest reinforcing bar.

Crank. An offset bend in a reinforcing bar to facilitate two bars being lapped without clashing.

Cross-section. A drawing taken from a view at right angles to the length of the member showing its smaller dimensions.

Cuphead. A description associated with a bolt or rivet which has a rounded head.

Distribution steel. The secondary reinforcement placed at right angles to the main reinforcement in a reinforced concrete slab or wall.

Downpipe. A vertical pipe which collects rain water from the gutter and transports it to the drain.

Eaves filler. Pieces of asbestos shaped to close the corrugations in roof sheeting at eaves and valleys.

Fabrication. Preparing the structural steel members by cutting, drilling, notching, machining, welding, etc.

Flashing. Sheet lead, metal or felt used where a pitched roof, lintel, sill, etc., meet a wall, to prevent water penetration.

Grout. Neat cement slurry or a mix of cement and sand used to fill the gaps around holding down bolts and under base plates.

Gusset plate. A steel plate shaped to connect the members of a truss or girder.

Gutter. A horizontal section designed to collect rain water from a roof.

Hinge. A point on a structure at which a member can rotate slightly and where the bending moment is considered to be zero.

Holding down bolts. Bolts cast into and projecting from the concrete foundation to hold down a structural steel section. Often called HD bolts.

Kicker. A stub portion of a reinforced concrete member, usually about 75 mm high, which enables the member to be satisfactorily formed. It gives a convenient method of clamping the bottom of the shutters.

Lap. The length by which one reinforcing bar must overlap with another to give effective uninterrupted reinforcement.

Lattice. An open girder, beam or column built up from main members joined by diagonal members.

Link stirrup or binder. A comparatively small diameter reinforcing bar which encloses the larger bars placed normal to it, so forming a reinforcement cage. They are usually located in beams and columns.

Main beam. A beam which bears directly on to a wall or column and not on to another beam.

Pack or Packing. A steel plate used to fill a gap between two other plates.

Plan. A drawing showing a "bird's-eye" view of a member.

Portal frame. A frame consisting of two vertical members rigidly connected at the top by another member which may be horizontal, curved or sloping.

Purlin. A horizontal member spanning between roof trusses or frames, supporting either roof sheeting or rafters for slates or tiles.

Reaction. The upward resistance of a support, such as a wall or column, against the downward pressure of a loaded member.

Resident engineer. Often referred to as the RE, he is the official appointed on the behalf of the client to ensure that the contractor's engineering work complies with that specified.

Ridge capping. Pieces of material used at the ridge or apex of a sloped roof to prevent moisture penetration.

Screed. A layer of mortar (sand/cement) laid on a floor or roof slab either as a bed for the finish or to give sufficient falls to assist the drainage of rain water.

Secondary Beam. A beam supported by main beams.

Splice. A joint between two lengths of a stanchion or a beam.

Stanchion. A vertical load bearing member of relatively small cross section, made from structural steelwork.

Starter bar. A short reinforcing bar cast into a supporting member to which may be lapped the reinforcement in the member supported.

Tolerance. A dimension to allow for the difference between the exact dimension required and that obtainable by the method of construction employed.

Washer. A steel ring placed under a bolt head or nut.

Index

Details of other titles
in the M&E TECBOOK series are to be found
on the following pages.

For a full list of titles and priccs, write
for the FREE TECBOOK leaflet and/or the Macdonald
& Evans Technical Studies catalogue, available from
Department MP1, Macdonald & Evans Ltd.,
Estover, Plymouth PL6 7PZ

Accommodation Operations
COLIN DIX

This book covers the basic procedures of hotel reception departments, including reservations, billing and cashiering, and also looks at tours and groups, and the ways in which the receptionist can increase sales in a hotel. There is a chapter outlining the latest electronic systems being introduced into hotels, and discussing the direction and impact of likely future developments. The book is intended to be used by students preparing for Higher National Certificate and Diploma, TEC, HCIMA and City and Guilds 709 Hotel Reception examinations, as well as students studying for degrees in catering management. "A useful addition for students taking exams on front office and reception." *Caterer and Hotelkeeper.*

Biochemistry Level III
P.L. DAVIES

Biochemistry is a multidisciplinary subject and interacts with many traditionally understood fields of study such as physiology, cell biology, and organic and analytical chemistry. At the same time, however, it is a subject in its own right, with its own research techniques and points of emphasis. The author shows this by drawing together facts from other disciplines so that the student gradually discovers what the scope of biochemistry is and how its own special techniques are applied. The book starts by discussing subjects that are also covered by other disciplines, for example stereochemistry and energetics, and then proceeds to examine such topics as enzymes, metabolism and biochemical genetics. Although written primarily for the TEC Level III Unit, the book will also be of use to first-year undergraduates and other students who need to have a knowledge of biochemistry and of the part biochemistry plays in their own particular field of interest.

Cell Biology Level II
BRIAN BONNEY

This concisely written and well-illustrated text is prepared with a view to meeting the needs of the TEC Level II student specialising in Cell Biology. Like the TEC syllabuses themselves it represents a new departure in the exposition of its subject material. It presents a thorough and up-to-date discussion at the appropriate level of our knowledge of cell structure and function, while at the same time relating this to the techniques by which this knowledge is obtained. It therefore bridges the gap between those books on cell biology which are entirely theoretical and those which are written as laboratory manuals.

Construction Technology Level I
K. HARWOOD

This book covers the objectives of the standard TEC unit U75/073. From an introduction to the built environment, it proceeds to a consideration of the substructure and superstructure of buildings.

Electrical and Electronic Principles Level II
F. GOODALL & D.K. RISHTON

This book is specifically designed to cover the syllabus requirements of the TEC unit of the same name. Written in a clear, explanatory style, key concepts are explained simply and analogies taken from everyday experiences. Many worked examples are included and each chapter concludes with a series of self-assessment questions.

Electronics Level II
PETER BEARDS

This book covers the syllabus of the first TEC unit specifically concerned with electronics. It deals first with devices and then examines their applications. Semiconductor devices dominate modern electronic engineering and accordingly a study of semiconductor theory, semiconductor diodes and the junction transistor are given prominence. Thermionic valves are also described because valve theory is the basis of the cathode ray tube and valves are widely used in radio transmitters. Amplifiers (especially transistor amplifiers), waveform generation and binary logic circuits are also covered in some detail. The book is amply illustrated and includes many worked examples.

Mammalian Physiology Level II
PHYLLIS A. ALLEN

This book is designed specifically to cover TEC's Mammalian Physiology Level II syllabus, although nurses studying for examinations and students preparing for G.C.E. in Biology and Human Biology at "O" and "A" Level will also find it invaluable. Topics covered include the structure and functions of the tissues of the body, the circulation of body fluids, respiration, diet, digestion, the liver and homeostasis. The book is illustrated with many clear diagrams, and students and teachers will appreciate its lucid, concise and up-to-date approach.

Multiple Choice Questions in Electrical Principles for TEC Levels I, II and III
A. DAGGER

This book consists of 360 multiple choice questions specifically prepared for people studying or teaching Electrical Principles Units at Levels I, II or III of a TEC Certificate course in Electrical or Electronic Engineering. It will be particularly useful to students who wish to use the questions for individual practice, revision or self-assessment.

Organisation and Procedures in the Construction Industry
P.A. WARD

This book aims to provide the student with a clear understanding of the complex structure and workings of the construction industry. Although specifically designed to meet the requirements of the TEC Level I Unit, Organisation and Procedures, the text has been broadened to meet the needs of students participating in the Joint Institute of Building/ National Federation of Building Trades Employers Site Management Education and Training Scheme.

Site Studies (Production) Level IV
P.A. WARD

This book aims to provide an introduction to the complex field of construction management and to enable the reader to develop a clear understanding of the numerous production techniques which will assist him in his work. Case studies are included to help the student apply the information in the text to practical situations. It is designed primarily for TEC Higher Certificate courses but will also be most useful to students preparing for final professional examinations of the Institute of Building, or following degree courses in building with a management content.

Site Surveying and Levelling Level II
M.R. HART

This book covers the objectives of the standard TEC unit U75/056. It introduces students to basic principles and deals with linear, height and angular measurements as well as building surveys and setting out.

OTHER RELEVANT M & E PUBLICATIONS

Civil Engineering Construction
L.M. JAMES

This HANDBOOK describes the common techniques currently used in the basic areas of civil engineering. It is aimed principally at students working towards Higher National Certificate or Diploma (or their equivalents) and the new TEC examinations. The book will be especially useful to students on courses with elements of industrial training who should be able to relate the text to practical experience.

Concise Soil Mechanics
M.J. SMITH

This HANDBOOK meets the requirements of students taking professional, degree and Higher National Diploma/Certificate examinations in soil mechanics. The areas covered include classification procedures, standard definitions, theoretical aspects of permeability, consolidation, shear and pressure distribution, foundation design, slope stability, earth-retaining structures and soil stabilisation. "This 'must' for civil engineering undergraduates is well-named because few current publications contain such a succinct text without glaring omissions or reams of superfluous detail." *New Civil Engineer*

Encyclopaedia of Metallurgy and Materials
C.R. TOTTLE

This large-format book is based on the successful *Dictionary of Metallurgy,* by Dr. A.D. Merriman, and not only covers the fields of metals and metallurgy, but also deals in detail with materials science. The entries are complemented by a large number of illustrations. The text is aimed at a wide range of readers, from the non-technical to students and practising scientists and technologists, while many of the tables and charts will be of everyday use in design offices, materials stores and supervisors' offices.

Land Surveying
RAMSAY J. P. WILSON

This popular HANDBOOK is intended for those studying for professional examinations in surveying, architecture and town planning. It will also be of value to those studying for first-year degree courses in civil and structural engineering, geography and geology, and as a practical reference book for those involved with surveying projects. ". . . recommended as being comprehensive, readable, easy to understand, and very good value." *Building Technology and Management*

Properties of Materials
C.V.Y. CHONG

This HANDBOOK is designed for students of architecture, building, engineering and surveying preparing for examinations in Properties of Materials or Materials Science at degree, professional and TEC higher level, and provides an introduction to the properties and uses of the more common materials used in the construction industry.

Public Health Engineering Practice
L.B. ESCRITT

This latest edition of the previously entitled *Work of the Public Health Engineer,* is a completely revised two-volume version of a popular standard work.
Volume I, *Water Supply and Building Sanitation,* deals with sources of water supply, the quality and chemistry of water, reservoirs, the removal of solids, sterilisation, pumping, indoor water systems and the drainage of premises.
Volume II, *Sewerage and Sewage Disposal,* discusses surface-water sewerage, the construction of sewers and manholes, sewage pumping, river maintenance, the chemistry, treatment and disposal of sewage, siting and layout of sewage-treatment works, screening, the treatment and disposal of sewage sludge, percolating-filter treatment, activated-sludge systems and ancillary subjects. "The work represents a vast store of information on all aspects of the water technology side of Public Health Engineering and is strongly recommended as a work of reference to practising engineers and students alike." *Public Health Engineer*

Town and Country Planning Law
A.M. WILLIAMS

This HANDBOOK provides a summary of the law relating to town planning in concise and comprehensive terms. Topics covered include the history of planning legislation, its administration and structure, local planning and the process of obtaining planning permission. Enforcement procedures are also covered along with special forms of control such as the listed building. The financial aspects of planning are considered, particularly with reference to development land tax. Legislation taken into account includes the Community Land Act 1975 and the Local Government, Planning and Land (No. 2) Act 1980. Suitable for all degree and professional examination syllabuses which require a knowledge of the subject, including those in architecture, civil engineering and surveying as well as geography, law and town planning.